Wood Structure in Plant Biology and Ecology

Wood Structure in Plant Biology and Ecology

edited by

Pieter Baas
Giovanna Battipaglia
Veronica De Micco
Frederic Lens
Elisabeth Wheeler

IAWA Journal 34 (4), 2013

International Association of Wood Anatomists
c/o Naturalis Biodiversity Center, Leiden, The Netherlands

Library of Congress Cataloging-in-Publication Data are available from the Publisher,
Leiden, The Netherlands.

ISBN: 978 90 04 26559 2

Koninklijke Brill NV incorporates the imprints Brill, Global Oriental,
Hotei Publishing, IDC Publishers and Martinus Nijhoff Publishers.

Printed in The Netherlands.

Contents

The page numbers in the above Table of Contents refer to the bracketed page numbers in this volume.

Wood Structure in Plant Biology and Ecology

Pieter Baas[1], Giovanna Battipaglia[2], Veronica De Micco[3], Frederic Lens[1], and Elisabeth Wheeler[4] (editors)

[1]Naturalis Biodiversity Center, PO Box 9517Leiden, The Netherlands
[2]Dipartimento di Scienze e Tecnologie Ambientali, Biologiche e Farmaceutiche (Di.S.T.A.Bi.F), Seconda Università di Napoli, via Vivaldi 43, I-81100 Caserta, Italy
[3]Dipartimento di Agraria, Università degli Studi di Napoli Federico II, via Università 100, I-80055 Portici (NA), Italy
[4]Department of Forest Biomaterials, NCSU, Raleigh, USA

This special issue of the IAWA Journal contains original and review papers presented at the successful meeting of the International Association of Wood Anatomists (IAWA), convened in the framework of the International Symposium "Wood Structure in Plant Biology and Ecology", held in Naples from 17–20 April 2013 and organized by the University of Naples Federico II and the Second University of Naples on behalf of the COST-Action STReESS, 'Studying Tree Responses to extreme Events: a SynthesiS'.

Xylem of trees, shrubs, and also herbs plays a crucial role in plant biology and ecology (Baas & Miller 1985). The evolutionary fitness of each plant species depends to a large extent on the fine balance between its hydraulic efficiency and safety, its cost-effective biomechanical design, and its biological defense, all in equilibrium with specific requirements posed by the physical and biological conditions of its environment. Moreover, wood constitutes a historical archive of the changing environmental and climate conditions as well as episodic or incidental stresses to which an individual tree or shrub has been exposed. This in turn allows studying plant responses to environmental change in past and present, and hypothesizing future effects of climate change. Almost 350 years after the microscopic structure and functions of wood were first discovered and hotly debated by the early microscopists Malpighi, Grew and Van Leeuwenhoek, the study of functional and ecological wood anatomy enjoys a renaissance and plays a pivotal role in plant and ecosystem biology, global change research and the understanding of plant evolution (Groover & Cronk 2013).

The programme of the Naples meetings represented a very full and rich spectrum of all aspects of wood structure in plant biology and ecology. In a uniquely cooperative spirit, four international Journals (*Dendrochronologia*, *IAWA Journal*, *Trees*, and *Tree Physiology*) collaborated to publish as many of the high quality papers presented at the meeting as possible. *Dendrochronologia* will publish a selection of tree-ring studies; *Trees* and *Tree Physiology* will publish some of the papers with a more strictly

© International Association of Wood Anatomists, 2013
Published by Koninklijke Brill NV, Leiden

DOI 10.1163/22941932-00000028

(eco)physiological focus. The IAWA Journal is proud to be the first to dedicate a full special issue to the proceedings of this meeting, including papers with an anatomical focus.

The selected papers give a good cross section of current developments in the study of functional and ecological wood anatomy. Review papers address the hydraulic architecture of the earliest woody land plants (Strullu-Derrien *et al.*), the general and functionally crucial phenomenon of axial conduit widening (Anfodillo *et al.*), the hydraulic and biomechanical optimization in a very important timber species (Rosner), and cellular and subcellular changes in the cambium in response to environmental factors (Prislan *et al.*). New tools to study structural aspects of hydraulic functioning of woody plants are reviewed in a paper on high-resolution 3-D visualization of wood structure (Brodersen), and applications of the ROXAS software to vessel detection and vessel grouping (papers by Wegner *et al.* and von Arx *et al.*). About half of the papers represent a bouquet of case histories in wood studies as applied to plant biology and ecology: on the effects of fire on wood structure (De Micco *et al.*) and on the occurrence of intra-annual density fluctuations in response to late summer/early autumn rains (Novak *et al.*), both in Aleppo pine. Stojnic *et al.* report on plastic growth and anatomical responses in beech. Two papers not presented in Naples, but submitted in recent months to the IAWA Journal also fit the theme of this special issue: a paper on changing growth rhythms in a widespread deciduous tropical hardwood (Costa *et al.*) and an overview of wood anatomical variation in ten common arctic dwarf shrub species (Schweingruber *et al.*). Altogether an inspiring collection of contemporary applications of wood anatomy to better understand the ecophysiology of woody plants.

Apart from the normal print-run as IAWA Journal 34 (4), Brill is also publishing a trade edition of this special issue. Editors and Guest Editors (listed here alphabetically) wish to thank all authors and anonymous reviewers for submitting manuscripts, reviews and revisions under considerable time pressure. We also wish to acknowledge the financial and logistic support of the meeting by the Cost-Action STReESS and its very cooperative Chair, Dr. Ute Sass-Klaassen.

References

Anfodillo T, Petit G & Crivellaro A. 2013. Axial conduit widening in woody species: a still neglected anatomical pattern. IAWA J. 34: 352–364.

Baas P & Miller RB. 1985. Functional and ecological wood anatomy – some introductory comments. IAWA Bull. n.s. 6: 281–282 (in special issue dedicated to proceedings of the Martin H. Zimmermann memorial symposium).

Brodersen CR. 2013. Visualizing wood anatomy in three dimensions with high-resolution X-ray micro-tomography (μCT) – A review. IAWA J. 34: 408–424.

Costa MS, de Vasconcellos TJ, Barros CF & Callado CH. 2013, Does growth rhythm of a widespread species change in distinct growth sites? IAWA J. 34: 498–509.

De Micco V, Zalloni E, Balzano A &Battipaglia G. 2013. Fire influence on *Pinus halepensis*: wood responses close and far from scar. IAWA J. 34: 446–458.

Groover A & Cronk Q. 2013. From Nehemiah Grew to genomics: the emerging field of evo-devo research for woody plants. Int. J. Plant Sci. 17: 959–963.

Novak K, Saz Sánchez MA, Čufar K, Raventós J & de Luis M. 2013. Age, climate and intra-annual density fluctuations in *Pinus halepensis* in Spain. IAWA J. 34: 459–474.

Prislan P, Čufar K, Koch G, Schmitt U & Gričar J. 2013. Review of cellular and subcellular changes in cambium. IAWA J. 34: 391–407.

Rosner S. 2013. Hydraulic and biomechanical optimization in Norway spruce trunkwood – A review. IAWA J. 34: 365–390.

Schweingruber FH, Hellmann L, Tegel W, Braun S, Nievergelt D & Büntgen U. 2013. Evaluating the wood anatomical and dendroecological potential of arctic dwarf shrub communities. IAWA J. 34: 485–497.

Stojnic S, Sass-Klaassen U, Orlovic S, Matovic B & Eilmann B. 2013. Plastic growth response of European beech provenances to dry site conditions. IAWA J. 34: 475–484.

Strullu-Derrien C, Kenrick P, Badel E, Cochard H & Tafforeau P. 2013. An overview of the hydraulic systems in early land plants. IAWA J. 34: 333–351.

von Arx G, Kueffer C & Fonti P. 2013. Quantifying plasticity in vessel grouping – added value from the image analysis tool ROXAS. IAWA J. 34: 433–445.

Wegner L, von Arx G, Sass-Klaassen U & Eilmann B. 2013. ROXAS - an efficient and accurate tool to detect vessels in diffuse-porous species. IAWA J. 34: 425–432.

Citation

Chapters in this book should be cited as regular papers in IAWA Journal 34 (4), using the high page numbers indicated on the title pages.

BRILL

IAWA Journal 34 (4), 2013: 333–351

AN OVERVIEW OF THE HYDRAULIC SYSTEMS IN EARLY LAND PLANTS

Christine Strullu-Derrien[1,*], **Paul Kenrick**[1], **Eric Badel**[2,3], **Hervé Cochard**[2,3] and **Paul Tafforeau**[4]

[1]Department of Earth Sciences, The Natural History Museum, Cromwell Road, London SW7 5BD, United Kingdom
[2]INRA, UMR547 PIAF, 63100 Clermont-Ferrand, France
[3]Clermont Université, Université Blaise Pascal, UMR547 PIAF, 63000 Clermont-Ferrand, France
[4]European Synchrotron Radiation Facility, 6 rue Jules Horowitz, 38043 Grenoble cedex, France
*Corresponding author; e-mail: c.strullu-derrien@nhm.ac.uk

ABSTRACT

One of the key functions of wood is hydraulic conductivity, and the general physical properties controlling this are well characterized in living plants. Modern species capture only a fraction of the known diversity of wood, which is well preserved in a fossil record that extends back over 400 million years to the origin of the vascular plants. Early fossil woods are known to differ in many key respects from woods of modern gymnosperms (*e.g.*, tracheid size, secondary wall thickenings, lignin chemistry, cambium development) and recent discoveries are shedding new light on the earliest stages of wood evolution, raising questions about the performance of these systems and their functions. We provide an overview of the early fossil record focusing on tracheid morphology in the earliest primary and secondary xylem and on cambial development. The fossil record clearly shows that wood evolved in small stature plants prior to the evolution of a distinctive leaf-stem-root organography. The hydraulic properties of fossil woods cannot be measured directly, but with the development of mathematical models it is becoming increasingly feasible to make inferences and quantify performance, enabling comparison with modern woods. Perhaps the most difficult aspect of hydraulic conductance to quantify is the resistance of pits and other highly distinctive and unique secondary wall features in the earliest tracheids. New analytical methods, in particular X-ray synchrotron microtomography (PPC-SRµCT), open up the possibility of creating dynamic, three-dimensional models of permineralized woods facilitating the analysis of hydraulic and biomechanical properties.

Keywords: Wood, fossil, synchrotron, 3D model, permineralization, hydraulic properties.

INTRODUCTION

The evolution of the vascular system in plants was a key development in the history of life because of its fundamental role in water transport and, in many species, its ancillary function as a framework of structural support (Sperry 2003; Pittermann 2010; Lucas

© International Association of Wood Anatomists, 2013
Published by Koninklijke Brill NV, Leiden

DOI 10.1163/22941932-00000029

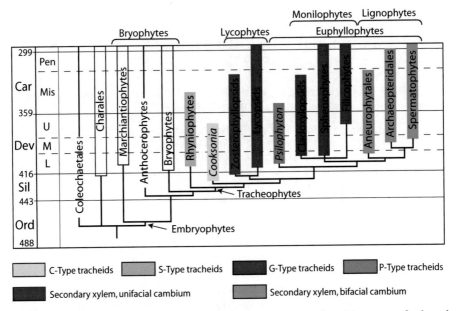

Figure 1. Simplified phylogenetic tree showing the minimum stratigraphic ranges of selected groups based on megafossils (bars) and their minimum implied range extensions (lines). Tracheid and secondary xylem types are shown on the figure. Ord = Ordovician; Sil = Silurian; Dev = Devonian, L = Lower, M = Middle, U = Upper; Car = Carboniferous. Mis = Mississippian, Pen = Pennsylvanian. Adapted from Kenrick & Crane (1997b).

et al. 2013). The vascular plants or tracheophytes are defined by the possession of this tissue system, the acquisition of which was essential to the evolution of their diverse forms, leading ultimately to their dominance of terrestrial ecosystems (Niklas 1997; Bateman *et al.* 1998; Labandeira 2005). The main constituents of the vascular system are phloem and xylem, but it is the latter that is more commonly encountered in the fossil record due to the resilience of its cellular components, which typically possess robust cell walls containing the polyphenolic polymer lignin (Boyce *et al.* 2004). Vascular tissues first appear in the fossil record in the lower part of the Devonian Period (410– 407 Myr) (Fig. 1) when terrestrial sediments containing fossil plants first became abundant (Gensel 2008; Kenrick *et al.* 2012). Research over the past 25 years has revealed some of the earliest stages in the evolution of the xylem (Stein 1993) and in particular has focused on the interpretation and documentation of its main component, the tracheid (Kenrick & Crane 1991; Edwards 1993; Friedman & Cook 2000; Sperry 2003). Several distinctive types of tracheid are now widely recognized, providing detailed information on the structure of the cell wall and insights into the early evolution of this important cell type.

The vascular system in the earliest tracheophytes was entirely primary, but the fossil record has also provided much information on the evolution of secondary vascular tissue (Cichan 1986; Cichan & Taylor 1990; Meyer-Berthaud *et al.* 2010; Spicer & Groover 2010; Gerrienne *et al.* 2011). Cambial activity in plants initially evolved independently at least twice in the Devonian Period (Kenrick & Crane 1997a,b) and perhaps on more occasions (Boyce 2010) (Fig. 1). In many extinct groups the cambium was unifacial, developing only secondary xylem, and it was unable to undergo anticlinal cell division (Taylor *et al*. 2009; Spicer & Groover 2010). This type of cambium produced the secondary xylem in the extinct tree cladoxylopsids of the Devonian Period and the lycopods and horsetails of the Carboniferous Period (Niklas 1997; Stein *et al*. 2007; Meyer-Berthaud *et al*. 2010). These trees generally produced rather little wood, and the vascular cylinder had limited capacity to increase in volume while maintaining the integrity of the cambium (Cichan 1986). The more familiar bifacial cambium gives rise to both secondary xylem and secondary phloem, and it has the capacity to undergo anticlinal cell division (Donoghue 2005). This form of cambium evolved during the Devonian Period in the lineage leading to gymnosperms and angiosperms (Hilton & Bateman 2006). New information from the early fossil record is providing further tantalizing insights that are promising to unravel the sequence of acquisition of characteristics that led to the evolution of secondary xylem (Gerrienne *et al*. 2011).

Our growing knowledge of the early evolution of vascular tissues derived from the careful study of fossils raises questions about the performance of these systems and their functional roles in the early development of vascular plants. Recent advances in our understanding of the hydraulics of modern woods (Hacke *et al*. 2004; Pittermann *et al*. 2006; Pittermann 2010) and the development and application of mathematical methods to infer the hydraulic properties of fossil woods (Wilson *et al*. 2008; Wilson & Knoll 2010; Wilson & Fischer 2011) mean that we now have many of the tools that we need to begin to investigate the hydraulic characteristics of the earliest vascular systems in plants. Early fossil wood is typically permineralized, and various methods that have been developed to investigate and characterize its structure include the preparation of sections for light microscopy and scanning electron microscopy (Jones & Rowe 1999). These are invasive, and they are frequently also destructive. The development of X-ray computed tomography, in particular the use of high-resolution tools such as synchrotrons, provides an efficient non-destructive alternative (*e.g*., Friis *et al*. 2007). Here we give a succinct overview of the earliest fossil record of primary and secondary xylem and its cellular components and an introduction to recent research on its hydraulic properties. We show how synchrotron microtomography can be used to investigate the hydraulic properties of the earliest wood.

THE EARLIEST FOSSIL EVIDENCE OF VASCULAR ELEMENTS AND THEIR HYDRAULIC STRUCTURES

Xylem and tracheid cell structure are frequently well preserved in the fossil record enabling both biomechanical properties and hydraulic efficiency to be estimated, furthering our understanding of the functional evolution of wood (Niklas 1997; Sperry 2003; Pittermann 2010). In fossils of the early part of the Devonian period the vascular system typically is permineralized in a variety of minerals including pyrite and its oxidation products (Kenrick 1999) (Fig. 2c–e), more rarely silicates (Channing & Edwards 2009), calcium/magnesium carbonates (Hartman & Banks 1980), or is simply preserved as

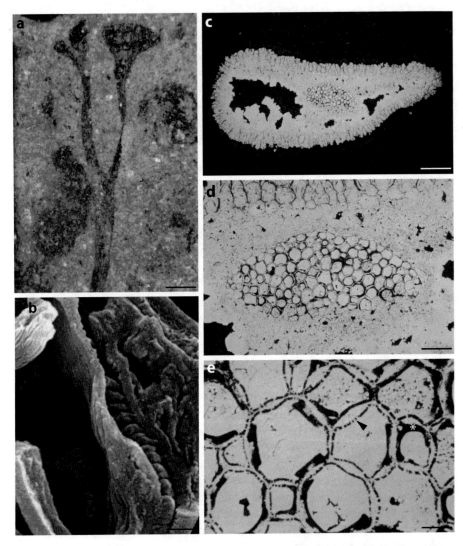

Figure 2. – **a**: *Cooksonia pertoni*: the earliest tracheophyte (Herefordshire, England; specimen nr. V 58010 Lang collection, Natural History Museum, London). Scale bar = 0.15 mm. – **b**: Coalified tracheid in *Cooksonia pertoni* (C-type tracheid). Note the thick coalified wall and mineral infill of lumen with grooves marking positions of secondary thickenings (From Edwards, New Phytologist, 1993; with courtesy). Scale bar = 0.8 μm. – **c–e**: Highly polished transverse section through a pyritized axis of *Gosslingia breconensis*. – **c**: Whole axis showing a central elliptical xylem surrounded by an amorphous acellular area of pyrite (white) and an outer cellular area. Scale bar = 250 μm. – **d**: Higher magnification of the xylem strand. Note the presence of wall pyrite between the coalified (black) walls of adjacent cells. Scale bar = 100 μm. – **e**: Transverse section through the cells at the edge of the xylem (G-type tracheid) showing components of the tracheid wall. Note the continuous organic wall of the annular or spiral wall sculpture (*) and the broken organic wall that represents a perforate wall (arrowhead) lying between the wall thickenings. Scale bar = 25 μm.

carbonized or charcoalified cells without associated mineralization (Edwards *et al.* 1992; Edwards 1993) (Fig. 2a, b). Interpreting cell wall structure is not straightforward, and the effects of decay and mineralization need to be critically evaluated when reconstructing tracheid characteristics and cell wall components (Kenrick & Edwards 1988; Kenrick & Crane 1991) (Fig. 3g–l). Some of the earliest plants possessed conducting systems comprising cells without the wall thickenings that characterize tracheids (Edwards 1993). By this we mean completely lacking helical or annular bars and pitting in the cell wall. One of the exceptionally well-preserved silicified plants from the 407-million-year-old Rhynie Chert (*Aglaophyton major*) possessed a conducting system that strongly resembles the basic organization of leptoids (specialized food-conducting cells) and hydroids (specialized water-conducting cells) observed in some of the larger modern mosses (Edwards 1993), but phylogenetic analysis shows that this fossil is more closely related to the vascular plants than to bryophytes (Kenrick & Crane 1997a, b). Tracheids may therefore have evolved from hydroid-like antecedents.

Early fossil vascular plants (Fig. 3a–f) possessed types of tracheid (Fig. 3g–l) that differ in significant ways from those of their modern relatives. One distinctive form is the S-type tracheid (Kenrick *et al.* 1991). This cell has large helical thickenings with a spongy interior (Fig. 3g, j). The lumen-facing surface is lined with a thin microperforate wall. Perforations within the wall measure c. 40 nm to 200 nm in diameter with a density of c. 16 μm^{-2}. The S-type tracheid has been observed in several stem group vascular plants (*Rhynia gwynne-vaughanii, Sennicaulis hippocrepiformis, Huvenia kleui, Stockmansella* sp.) (Fig. 1; 3a, d) in both their sporophyte and probably also their gametophyte generations and in two different forms of permineralization (*i.e.*, pyrite, silicates; Kenrick & Crane 1991). The G-type tracheid is a second distinctive form that is widespread in stem group lycopods (Fig. 1; 3b, e). The cell is characterized by annular or helical thickenings with some cross connections (Fig. 3h, k). Typically, a distinctive perforate sheet of material occupies the cell wall between the thickenings (Kenrick & Edwards 1988), but the thickenings themselves are known to be perforate in one species (Wang *et al.* 2003). Perforations in this layer are typically an order of magnitude larger than those in the S-type cell and of less regular shape (Kenrick & Edwards 1988). The wall thickenings of both S-type and G-type cells are essentially helical or annular, but the extensive development of cross connections between bars can lead to simple reticulate pitting in the G-type cell (Kenrick & Crane 1997a, b).

Bordered pitting developed in other tracheid types. This was an early innovation in vascular plants that evolved independently at least twice: once in lycophytes and once in euphyllophytes (Kenrick & Crane 1997a, b). Bordered pitting characterizes the P-type tracheid, which is common to many basal euphyllophytes (Fig. 1; 3c, f). Pitting is mostly of the scalariform type, and a distinctive feature is the presence of an additional perforate sheet of wall material extending over the pit apertures (Fig. 3i, l). The perforations in this sheet are distributed either in one of two transverse rows or less regularly in a reticulum (Hartman & Banks 1980; Kenrick & Crane 1997a, b). Within the lycophytes, a broadly similar pit construction is seen in the fossil *Minarodendron*, but the scalariform bars are much more elongate (Li 1990). These are slightly different to the P-type cell. In *Minarodendron*, the additional perforate sheet of wall material

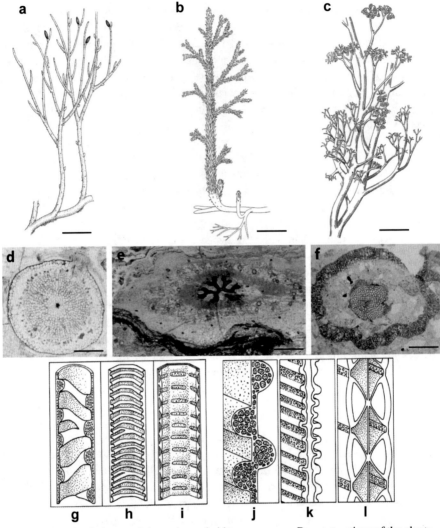

Figure 3. Early land plants and the main tracheid types. — **a–c**: Reconstructions of the plants. –
a: *Rhynia gwynne-vaughanii* (from Kenrick & Crane 1997a). Scale bar = 1 cm. – **b**: *Asteroxylon mackiei* (from Kidston & Lang 1920). Scale bar = 1.5 cm. – **c**: *Psilophyton dawsonii* (from
Kenrick & Crane 1997a). Scale bar = 1 cm. — **d–f**: Transverse sections of the axes. – **d**: *Rhynia gwynne-vaughanii* (slide nr. 3133, Scott collection, Natural History Museum, London). Scale
bar = 0.6 mm. – **e**: *Asteroxylon mackiei* (slide nr. 1015, Lang collection, Natural History Museum,
London). Scale bar = 2.5 mm. – **f**: *Psilophyton dawsonii* (slide nr. OH55 x.2, Florin collection,
Naturhistoriska riksmuseet, Stockholm). Scale bar = 0.6 mm. — **g–l**: Diversity of tracheids in
early land plants (median longitudinal section through the cells; the basal and proximal end walls
are not shown; cells are 20–40 μm in diameter). – **g**: S-type tracheid. – **h**: G-type tracheid. –
i: P-type tracheid. – **j**: details of S-type cell wall showing 'spongy' interior to thickenings and
distribution of perforations in thin lumen-facing layer. – **k**: details of G-type cell wall showing
perforations distributed between thickenings. – **l**: details of P-type cell wall showing pit chambers
and layer with perforations that extends over pit apertures.

is attached to the pit aperture crossing the pit chamber, whereas in the P-type cell of *Psilophyton* it is positioned within the pit and closer to the pit-closing membrane. This additional perforate sheet of secondary wall material associated with pitting does not occur in modern groups, and it appears to be relatively short-lived in geological terms. Within euphyllophytes and many lycopods it is soon supplanted in the Devonian Period by tracheids with more conventional bordered pits. The feature persists in some arborescent lycopods (Cichan *et al.* 1981) into the Carboniferous Period. The functional significance of the additional perforate sheet crossing pit chambers is unknown, but we suggest that it would have increased the hydraulic resistance of the cell wall, reducing the risk of cavitation.

The three types of early vascular element discussed above are widely recognized and have been characterized in some detail, but other variants are known to exist (*e.g.*, C-type; Edwards 1993) (Fig. 2a, b). By the middle part of the Devonian Period tracheids with more conventional bordered pits without the additional perforate sheets of wall material had evolved (*e.g.*, *Leclercqia*; Grierson 1976). In early vascular plants, the metaxylem is generally thought to be composed primarily of one or other of the main tracheid types. Thus, at any given point in a plant, the entire metaxylem would be composed of either the S-type, G-type or P-type tracheids. The assumption that one tracheid type would characterize a whole plant has recently been refuted with the discovery of G-type tracheids at one level and a pitted type at a more distal level within the same individual (Edwards *et al.* 2006). Thus, tracheid wall structure may vary quite significantly at different levels within an individual.

The major features of the secondary walls of the earliest tracheids (*i.e.*, presence or absence of thickenings and pitting) are distinctive and readily recognizable. One other common characteristic is the distribution pattern of organics and minerals observed within the permineralized cell walls. The significance of this needs to be interpreted carefully and with reference to the taphonomic processes involved (Kenrick & Crane 1991). In certain forms of permineralization, notably pyrite and its oxide derivatives, the partitioning of mineral and organics (Fig. 2e) most likely reflects the distribution of resistant polyphenolics (*i.e.*, lignin) within the basic cellulose framework of the cell wall (Kenrick & Edwards 1988). Under this interpretation, the coalified parts of the wall are decay resistant and reflect the original pattern of lignification, whereas the mineralized parts represent areas that underwent substantial decay and are therefore likely to have been weakly lignified or non-lignified. This hypothesis has been tested through a detailed analysis of tracheid development and subsequent decay in the living lycopod *Huperzia*. Friedman and Cook (2000) showed that patterns of lignification in the tracheids of *Huperzia* broadly reflect those hypothesized in the early fossil tracheids.

The fossil record shows that lignification of the cell wall in the earliest vascular plants was most well developed on the inner lumen-facing surface, but the thickness of this layer varied among cell types (Friedman & Cook 2000). In all cells that possess this feature, the inner lumen-facing lignified layer was perforate (Fig. 3j–l). The perforations were most numerous and smallest in the thin lignified wall of the S-type cell (Fig. 3j). Here, their size and distribution indicate that they might be plasmodesmata derived, because they share similarities to pores in the water-conducting cells of some living

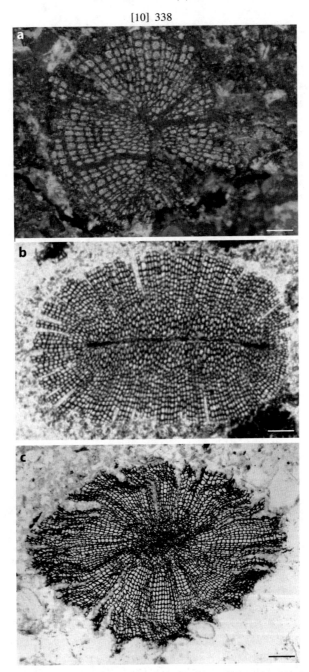

Figure 4. The three oldest (Early Devonian) euphyllophytes exhibiting wood. Transverse section of the xylem. – **a**: The plant from Châteaupanne (Armorican Massif, France). Scale bar =
200 μm. – **b**: The plant from New Brunswick (Canada) (photo P. Gensel, with courtesy). Scale
bar = 300 μm. – **c**: *Franhueberia gerrienni* from Gaspé (Canada) (photo M. Tomescu, with
courtesy). Scale bar = 200 μm.

liverworts (*e.g.*, Ligrone & Duckett 1996). The perforations were larger and typically more restricted in distribution in the thicker walls of the G-type tracheids (Fig. 3k), and they reached their most developed form in the bordered pit (Fig. 3l). As with bordered pits, from a functional perspective it is probable that the various forms of perforations in the walls of these early tracheids were channels enabling the flow of fluids through an otherwise impermeable wall layer.

THE EARLIEST EVIDENCE OF WOOD

The first forest ecosystems evolved by the middle part of the Devonian Period, and they were populated by arborescent plants belonging to several major lineages (Bateman *et al*. 1998; Gensel & Edwards 2001; Stein *et al*. 2007; Gensel 2008; Meyer-Berthaud *et al*. 2010; Cornet *et al*. 2012; Stein *et al*. 2012; Giesen & Berry 2013). Arborescence is known to have evolved in plants independently in many different groups, and a variety of biomechanical strategies were employed (Mosbrugger 1990; Niklas 1997; Rothwell *et al*. 2008; Pittermann 2010). In gymnosperms, the evolution of wood was key to the development of shrubs and trees, whereas it played a lesser role in the evolution of arborescence in several other groups of plants (Niklas 1997; Donoghue 2005; Meyer-Berthaud *et al*. 2010). The wood of early gymnosperms was complex, derived from a bifacial cambium which, through periclinal cell division, gave rise to secondary xylem towards the centre and secondary phloem towards the outside. The secondary xylem contained both tracheids and rays. The bifacial cambium was also able to undergo anticlinal cell division to accommodate increasing girth (Niklas 1997; Spicer & Groover 2010). Other early woods in the extinct arborescent cladoxylopsids, lycophytes and sphenophytes differed in some important features. With the possible exception of *Sphenophyllum* (Eggert & Gaunt 1973; Cichan & Taylor 1982), their cambia are generally thought to have been unifacial, producing only secondary xylem (Niklas 1997; Spicer & Groover 2010). Furthermore, in many but not all early woody plants the cells of the unifacial cambium were unable to divide anticlinally to produce new cambial initials. Thus, the cambium in these plants had limited capacity to increase in circumference and retain its integrity as girth increased (Donoghue 2005; Spicer & Groover 2010). Early fossils therefore show that the suite of characteristics that comprise wood in modern gymnosperms assembled over a period of time, with the capacity to produce secondary xylem appearing prior to the evolution of secondary phloem, and periclinal cell division of cambial initials appearing before the ability to sustain anticlinal divisions indefinitely. Recent research on fossils from the early part of the Devonian Period is beginning to shed further light on the earliest stages in the evolution of wood.

The earliest vascular plants possessed entirely primary growth, but among euphyllophytes one occasionally sees short files of radially aligned xylem, giving the impression of secondary growth. This was observed in Lower Devonian fossils such as *Psilophyton dawsonii* (Banks *et al*. 1975) and *Psilophyton crenulatum* (Doran 1980), which were leafless plants of small stature and rhizomatous growth without a well-developed root system (Fig. 3c). In their larger axes only, both plants had xylem aligned in short radial files, but there is no evidence of rays, anticlinal cell division, or secondary phloem.

So, in these instances, it seems likely that alignment of xylem cells took place through periclinal cell divisions during primary growth.

The earliest evidence of secondary xylem in the fossil record also comes from euphyllophytes. Three occurrences have been reported recently from the Lower Devonian (Hoffman & Tomescu 2013), but the fossils are small and fragmentary, so much remains to be learned about their overall morphology. The earliest and most complete is the plant from Châteaupanne, Armorican Massif, France (late Pragian to earliest Emsian, c. 407 Ma), which is thought to closely resemble *Psilophyton* in overall morphology and mode of branching (Strullu-Derrien *et al.* 2010; Gerrienne *et al.* 2011; Strullu-Derrien *et al.*, submitted). Anatomical resemblances extend to the overall shape of the xylem in transverse section (Fig. 4a), the development of the primary xylem (centrarch), and the P-type tracheids (Hartman & Banks 1980; Kenrick & Crane 1997a,b). The plant from Châteaupanne differs from *Psilophyton* in the presence of secondary xylem (Gerrienne *et al.* 2011; Strullu-Derrien *et al.*, submitted). The xylem cells are aligned in radial files but unlike *Psilophyton* there is evidence of anticlinal cell division within the cell files (Fig. 4a; 5b, c), and possibly also the remains of a cambial layer (Gerrienne *et al.* 2011, fig. 4E). Rays are present (Fig. 4a), and they are probably uniseriate (Strullu-Derrien *et al.* submitted). The rays are rare and their form and size are very variable (Gerrienne *et al.* 2011, fig. 1A, F, G and fig. S2, S3 Supporting online material). Another distinctive feature, common in places, is the presence of tracheids with smaller radial diameter at the periphery of the wood. This configuration of cells might indicate the presence of a growth layer boundary within the wood or near its circumference or perhaps differences in the pattern of divisions of the fusiform initials along the cambial layer.

Secondary xylem was also recently reported in an unnamed fragment of plant axis from New Brunswick, Canada (Late Emsian; c. 397 Ma). The plant had a protoxylem and a metaxylem that was oval-elongate in form (Fig. 4b). The secondary xylem was composed of P-type tracheids and contained relatively numerous rays (Fig. 4b and Gerrienne *et al.* 2011, fig. 1D). A third taxon, *Franhueberia gerriennei* (Gaspé, Canada; Late Emsian; c. 397 Ma) (Fig. 4c), is known from a short length of a rather distorted permineralized axis that also possessed secondary xylem of P-type tracheids. The rays are relatively numerous and they are regularly distributed (Hoffman & Tomescu 2013). Common features that can be observed or inferred in the three earliest woody plants known include 1) presence of a cambium with both tracheid and ray initials, 2) development of secondary growth in small axes, and 3) P-type tracheids. In all three cases absence of secondary phloem in the fossilized remains of the plants could reflect either true absence or inadequate preservation, so we cannot confidently define the cambium as either unifacial or bifacial. Although these fossils are fragmentary they provide evidence for a type of or approximation of secondary growth in small stature plants prior to the evolution of a distinctive leaf-stem-root organography (Hoffman & Tomescu 2013).

ESTIMATING THE HYDRAULIC PROPERTIES OF EARLY FOSSIL XYLEM

Our focus here is the hydraulic properties of early fossil xylem, which cannot be measured directly, but which can be estimated or modelled using approaches developed on living plants (Pittermann 2010). The hydraulic efficiency of xylem is proportional to the conduit diameter raised to the fourth power (Hagen-Poiseuille relation), so wide conduits are advantageous over narrow channels (Tyree & Zimmermann 2002). The consequences of this relationship on the early evolution of xylem was first clearly demonstrated by Niklas (1985), who documented an 18-fold increase in maximum tracheid diameter during the initial diversification of vascular plants in the Devonian Period. Here, the evolution of greater hydraulic efficiency was undoubtedly related to the great increases in plant size and complexity that characterized this period of time. Using an equation derived from the Hagen-Poiseuille relation, Cichan (1986) was the first to attempt to quantify specific conductance in the secondary xylem of fossil plants with the aim of comparing values obtained to those in modern groups. Tree forms and lianas were chosen from several major extinct groups of pteridophytes (Calamitaceae, Lepidodendraceae, Sphenophyllaceae) and gymnosperms (Cordaitaceae, Medullosaceae) of the Carboniferous Period. Results indicated that conductance in some of the most ancient woody groups was comparable to that in living plants. Highly effective conducting tissues had therefore developed relatively early in plant evolution. Also of interest is that some of the general relationships between wood anatomy, growth habit, and ecology known in living plants appeared to hold for these early fossils.

Most Medullosans had an unusual and distinctive habit with slender stems that bore massive pinnately compound leaves. The vascular systems contained exceptionally wide tracheids, which is consistent with the inferred high evapotranspiration demands of the leaves (Wilson *et al*. 2008). The remarkably high inferred conductivities of the tracheids of *Lyginopteris*, *Callistophyton*, and especially *Medullosa* are similar to those of some vessel-bearing angiosperms, and the vascular anatomy indicates that they played little or no structural role in supporting stems. For some species this is suggestive of a semi-self-supporting vine-like or perhaps scandent habit (Wilson & Knoll 2010). It also indicates that these pteridosperms would not have fared well in seasonally arid or frost-prone environments, which is consistent with their inferred ecological setting as components of tropical floodplains floras (Wilson & Knoll 2010).

The model used by Cichan (1986) treated the xylem in the stem as a single set of unobstructed pipes extending from roots to leaves and thus overestimated flow volumes through stems (Wilson *et al*. 2008). In other words, this simplification does not take into account the fact that tracheids have a finite length, typically in the range of about 0.5 mm to 4 mm in living conifers, and that flow between cells occurs through pits in the walls that offer significant additional resistance, reducing actual conductivity by well over 40% of that predicted by the Hagen-Poiseuille relation. Furthermore, the degree of this resistance varies with pit morphology, the porosity of the wall within the pit, and the size, number and distribution of pits, and values for this are not well characterized for living pteridophytes, many gymnosperms and their extinct relatives (Pittermann 2010). In general, we would expect the resistance of pits in the earliest fossils to be higher than those of modern plants due to the presence of an additional lignified perforate wall layer. Wilson *et al* (2008) took the estimation of hydraulic conductance in early vascular plants a step further by developing a model for water transport in xylem conduits that accommodated resistance to flow from pits and pit

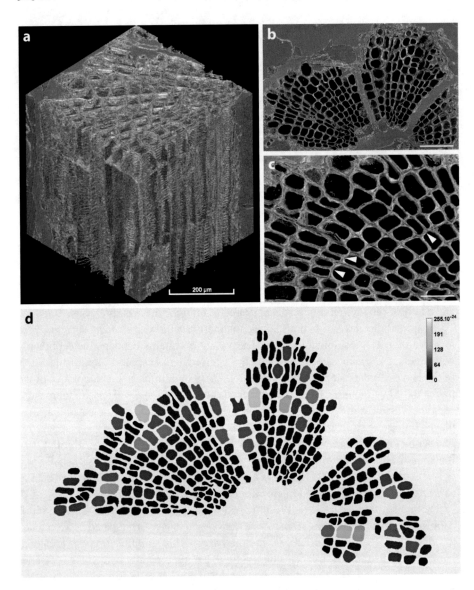

Figure 5. High-resolution Propagation Phase Contrast X-ray Synchrotron Microtomography (PPC-SRμCT) of 407 million year old fossil wood preserved in the mineral pyrite (FeS2) (Specimen CSD-07F-01, Université d'Angers, France). – **a**: Three-dimensional representation of part of pyritized axis of the plant from Châteaupanne, which has been virtually trimmed to a cubic volume. Xylem tracheids are visible in longitudinal, radial and transverse sections. – **b**: Transverse section extracted from the PPC-SRμCT part of xylem embedded in shale matrix; the mineral pyrite has been virtually dissected out leaving behind the organic framework of the tracheid cell walls. Scale bar = 250 μm. – **c**: Higher magnification of a transverse section extracted from the PPC-SRμCT part of xylem. At several places two xylem cell rows emanate from a single row (arrows). Scale bar = 50 μm. – **d**: Hagen-Poiseuille law prediction of lumens conductance based on a transverse section of part of xylem of the plant from Châteaupanne extracted from

membranes, and they applied their model to compare hydraulic resistance in two of the extinct gymnosperms modelled by Cichan (1986) (Cordaitaceae, Medullosaceae) and living *Pinus*. One important distinction between the tracheids of these gymnosperms is that pit membranes in *Pinus* are differentiated into a torus and margo, whereas in Cordaitaceae and Medullosaceae the pit membranes are homogeneous. Also, the second-ary xylem tracheids of extinct *Medullosa* are among the widest (commonly > 200 µm) and longest (> 17.5 mm) known in seed plants (Andrews 1940). Results showed that medullosan tracheids had the capacity to transport water at flow rates more compara-ble to those of angiosperm vessels than to those characteristic of modern conifers and their ancient relatives (Cordaitaceae). Furthermore, results indicated that the tracheids operated at significant risk of embolism and implosion, making this plant unlikely to survive significant water stress. These observations are consistent with the general paleo-ecological interpretation of some medullosans as large-leaved lianas growing on tropi-cal floodplains (DiMichele *et al.* 2006).

This general approach was extended recently to make a broad comparison of hydrau-lic conductance across seed plants, based on a sample of 22 living and extinct species (Wilson & Knoll 2010). A morphometric approach was used comparing across groups the key factors governing hydraulic resistance: cell length, cell diameter, and pit resist-ance. Results showed that extinct coniferophytes fall within the range of living conifers. The efficiency of torus-margo pitting could be matched for species with homogeneous pit membranes by increasing pit area. Living cycads, extinct cycadeoids and *Ginkgo* overlapped with both conifers and vesselless angiosperms. However, three Palaeozoic seed plants (*Lyginopteris, Callistophyton, Medullosa*) stood out as occupying a unique portion of the morphospace. These extinct species therefore evolved a combination of tracheid morphologies and xylem architectures that lay outside the range observable in living gymnosperms and angiosperms.

Much less is known about the comparative hydraulics of living ferns, horsetails, lyco-pods and their extinct relatives. Furthermore, in stem group vascular plants, tracheids differ in significant details of wall structure (*e.g.*, S-type, G-type, P-type) to modern forms, further complicating the modelling of their hydraulic properties. Wilson and Fischer (2011) modelled the hydraulic resistance of the tracheid cell wall of the early fossil *Asteroxylon* using a simple scalariform pit model. However, the tracheids in *Asteroxylon* are of the G-type (Kenrick & Crane 1991); the pitting is not scalariform but basically annular developing into reticulate over part. Also, the model did not consider

---

←

PPC-SRµCT. In this section, the mineral pyrite has been virtually dissected out leaving behind the organic framework of the tracheid cell walls. The color intensity of each xylem conduit refers to its conductance performance k_i. The individual conductivities k_i of cells were computed using the hydraulic diameter approximation that was proposed by Sisavath *et al.* (2001) for undefined cross sections of conduits:

$$k_i = \frac{1}{32} \, \mu D_{H_i}^{2} \, A_i$$

where A_i is the cross-sectional area of the conduit and D_{H_i} is its hydraulic diameter (Strullu-Derrien *et al.*, submitted).

the effects of the presence of the additional perforate sheet of secondary wall material extending between annular bars, which significantly restricts the effective porosity of the tracheid wall. The flow from one cell to another through the wall of a G-type cell is effectively defined by the area of perforation in the secondary wall. The calculated values therefore underestimate the hydraulic resistance of this tracheid type. Despite difficulties such as these, the development of more sophisticated models incorporating pit architecture and distribution hold promise in furthering our understanding of the hydraulic properties of early wood.

In addition to the theoretical considerations of modeling hydraulic conductance outlined above, there are technical and methodological issues concerning the characterization of both primary and secondary xylem in fossil plants. One set of issues relates to the imaging and measurement of tracheid cell walls. Typically, this is done by physical preparation of permineralized fossils to make transverse, radial and tangential sections (Jones & Rowe 1999). Measurements and imaging of tracheids would involve light microscopy. The method is destructive (*i.e.*, results in loss of materials) and usually the number of preparations that one can make is rather limited. Scanning electron microscopy is used typically to develop detailed reconstructions of pit structure. Recently we showed how Synchotron microtomography can be used to document the structure of the earliest wood and to collect measurements to perform calculations on hydraulic conductivity (Fig. 5a–d and Strullu-Derrien *et al.* submitted).

THE POTENTIAL OF SYNCHROTRON MICROTOMOGRAPHY

Synchotron microtomography (Feist *et al.* 2005; Lak *et al.* 2008; Tafforeau & Smith 2008) provides a new approach to investigating the structure and the hydraulic properties of early fossil wood (Strullu-Derrien *et al.*, submitted). The method is (i) non-invasive and non-destructive, (ii) enables the visualization of wood volumes in three dimensions, and (iii) allows dynamic virtual dissection in any number of transverse, radial, and tangential sections to explore properties at the cellular level (Fig. 5a). For example, we were able to trace rays through a block of xylem and visualize in longitudinal tangential section the interface with xylem tracheids and rays. Adjacent tracheids have a double wall (consisting of the secondary walls of each tracheid whereas the interface with the ray shows only a single wall (tracheid wall) as cells within the ray are not preserved (Strullu-Derrien *et al.*, submitted). Also virtual 3D histological sections (Fig. 5b,c) were generated from the reconstructed 3D volume, and used to estimate the hydraulic conductivity. Shape parameters (A_l and p_i) of the cell lumens were measured in transverse section using the ImageJ software (Rasband 2012) (Fig 5d) (Strullu-Derrien *et al.*, submitted).

With standard paleobotanical methods, such observation and measurement require sectioning in three orientations (transverse, radial and tangential) and the using of etching agents, and they are destructive. Much early fossil wood is permineralized in pyrite (FeS_2), which may become partially or completely oxidized (Kenrick 1999). The permineralization process results in mineralization that is believed to reflect the distribution of decay-resistant organics (*i.e.*, lignin) within the original cellulose matrix of the cell wall (Kenrick & Crane 1991; Edwards 1993). Synchrotron microtomography makes possible the virtual removal of the mineral to reveal the coalified remains of the cell walls, thereby eliminating the need for chemical etching (Fig. 5b, c). Imaging is based on the attenuation of X-rays as they pass through the object, which is related to its density and chemical composition. It is possible that this approach would also work for woods preserved in other common minerals, including calcium, magnesium, and iron carbonates, and silicates. Tomography is dependent on density contrasts and it has to be noted that results that are very informative for fossils preserved in pyrite may be less so for fossils permineralized in calcite and silica. The method will probably be most effective where mineral replacement of the organic remains of the cell walls is incomplete. The volume of wood we reconstructed is small (Fig 5a), but in principle larger volumes can be imaged, allowing the accurate measurement of tracheid length and also an assessment of the distribution and density of pits within tracheid cell walls; however, there is going to be a trade-off between volume imaged and resolution. Larger volumes generally result in lower resolution. Our results show that the method works for early wood preserved in pyrite; it enables the dynamic 3D imaging of the wood, and it facilitates the analysis of its hydraulic and biomechanical properties in a non-invasive and non-destructive way.

CONCLUSION

Understanding the early evolution of hydraulic conductivity in wood requires a detailed documentation of early fossils. The xylem in fossils is often well-preserved providing information on the structure of the tracheids and the general form and composition of the vascular system. Recent studies show that early fossil woods differ in many key respects to those of modern gymnosperms (*e.g.*, tracheid size, secondary wall thickenings, lignin chemistry, cambium development).

Hydraulic conductivity cannot be measured directly in fossils, but estimates can be obtained by measuring various properties of the fossilized water-conducting cells and using these as values in increasingly sophisticated biophysical models. This approach holds great promise in furthering our understanding of the hydraulic properties of early wood however modelling pit resistance in early fossils remains challenging.

Synchrotron microtomography is a flexible new and non-invasive tool for the study of permineralized woods that enables their dynamic virtual dissection. The method is effective for woods preserved in pyrite (FeS_2), and might also prove effective for woods preserved in other common minerals. The method facilitates the collection of measurements needed to calculate hydraulic conductivity, in a non-destructive way and for very small samples, which is a clear advantage.

The fossil record shows that wood evolved in small stature plants prior to the evolution of a distinctive leaf-stem-root organography. It also demonstrates that the suite of characteristics that comprise wood in modern gymnosperms assembled over a period of time. Results are beginning to show combinations of features in fossil woods that are outside of the range observed in modern plants. Knowledge of the hydraulic and the biomechanical properties of fossils woods can also help inform on the growth habit and ecology of extinct plants.

ACKNOWLEDGEMENTS

The authors thank Pieter Baas and Elisabeth Wheeler, Editors-in-Chief of the IAWA Journal for their invitation to write this article. They thank Dianne Edwards, Patricia G. Gensel (who studied the plants from New Brunswick) and Alexandru M. Tomescu for permission to reproduce figures. C.S-D received financial support from the European Commission under the Marie Curie Intra-European Fellowship Programme FP7-People-2011-SYMBIONTS.

REFERENCES

Andrews HN. 1940. On the stelar anatomy of the pteridosperms with particular reference to the secondary wood. Ann. Missouri Bot. Gard. 27: 51–118.

Banks HP, Leclercq S & Hueber FM. 1975. Anatomy and morphology of *Psilophyton dawsonii*, sp.n. from the late Lower Devonian of Quebec (Gaspé), and Ontario, Canada. Palaeontograph. Amer. 48: 77–127.

Bateman RM, Crane PR, DiMichele WA, Kenrick P, Rowe NP, Speck T & Stein W. 1998. Early evolution of land plants: phylogeny, physiology, and ecology of the primary terrestrial radiation. Ann. Rev. Ecol. Syst. 29: 263–292.

Boyce CK. 2010. The evolution of plant development in a paleontological context. Curr. Opin. Pl. Bio. 13: 102–107.

Boyce CK, Zwieniecki MA, Cody GD, Jacobsen C, Wirick S, Knoll AH & Holbrook NM. 2004. Evolution of xylem lignification and hydrogel transport regulation. Proc. Natl Acad. Sci. USA 101: 17555–17558.

Channing A & Edwards D. 2009. Silicification of higher plants in geothermally influenced wetlands: Yellowstone as a Lower Devonian Rhynie analog. Palaios 24: 505–521.

Cichan MA. 1986. Conductance in the wood of selected Carboniferous plants. Paleobiol. 12: 302–310.

Cichan MA & Taylor TN. 1982. Vascular cambium development in *Sphenophyllum*: a Carboniferous arthrophyte. IAWA Bull. n.s. 3: 155–160.

Cichan MA & Taylor TN. 1990. Evolution of cambium in geologic time – a reappraisal. In: Iqbal M (ed.), The vascular cambium: 213–221. John Wiley & Sons, New York.

Cichan MA, Taylor TN & Smoot EL. 1981. The application of scanning electron microscopy in the characterization of Carboniferous lycopod wood. Scan. Elect. Microsc. 3: 197–201.

Cornet L, Gerrienne P, Meyer-Berthaud B & Prestianni C. 2012. A Middle Devonian *Callixylon* (Archaeopteridales) from Ronquières, Belgium. Rev. Palaeobot. Palynol. 183: 1–8.

DiMichele WA, Phillips TL & Pfefferkorn HW. 2006. Paleoecology of Late Paleozoic pteridosperms from tropical Euramerica. J. Torrey Bot. Soc. 133: 83–118.

Donoghue MJ. 2005. Key innovations, convergence, and success: macroevolutionary lessons from plant phylogeny. Paleobiol. 31: 77–93.

Doran JB. 1980. A new species of *Psilophyton* from the Lower Devonian of northern New Brunswick, Canada. Can. J. Bot. 58: 2241–2262.

Edwards D. 1993. Cells and tissues in the vegetative sporophytes of early land plants. New Phytol. 125: 225–247.

Edwards D, Davies KL & Axe L. 1992. A vascular conducting strand in the early land plant *Cooksonia*. Nature 357: 683–685.

Edwards D, Li CS & Raven JA. 2006. Tracheids in an early vascular plant: a tale of two branches. Bot. J. Linn. Soc. 150: 115–130.

Eggert DA & Gaunt DD. 1973. Phloem of *Sphenophyllum*. Amer. J. Bot. 60: 755–770.

Feist M, Liu J & Tafforeau P. 2005. New insights into Paleozoic charophyte morphology and phylogeny. Amer. J. Bot. 92: 1152–1160.

Friedman WE & Cook ME. 2000. The origin and early evolution of tracheids in vascular plants: integration of palaeobotanical and neobotanical data. Phil. Trans. Roy. Soc. London B355: 857–868.

Friis EM, Crane P, Pedersen KR, Bengston S, Donoghue PCJ, Grimm GW & Stampanoni M. 2007. Phase-contrast X-ray microtomography links Cretaceous seeds with Gnetales and Benettitales. Nature 450: 549–552.

Gensel PG. 2008. The earliest land plants. Ann. Rev. Ecol. Syst. 39: 459–477.

Gensel PG & Edwards D. 2001. Plants invade the land: evolutionary and environmental perspectives. Critical Moments and Perspectives in Earth History and Paleobiology Series. Columbia University Press, New York.

Gerrienne P, Gensel PG, Strullu-Derrien C, Lardeux H, Steemans P & Prestiani C. 2011. A simple type of wood in two Early Devonian plants. Science 333 (6044): 837.

Giesen P & Berry CM. 2013. Reconstruction and growth of the early tree *Calamophyton* (Pseudosporochnales, Cladoxylopsida) based on exceptionnally complete specimens from Lindlar, Germany (mid-Devonian): organic connection of *Calamophyton* branches and *Duisbergia* trunks. Int. J. Plant Sci. 174: 665–686.

Grierson JD. 1976. *Leclercqia complexa* (Lycopsida, Middle Devonian): its anatomy, and the interpretation of pyrite petrifactions. Amer. J. Bot. 63: 1184–1202.

Hacke UG, Sperry JS & Pittermann J. 2004. Analysis of circular bordered pit function. II. Gymnosperm tracheids with torus-margo pit membranes. Amer. J. Bot. 91: 386–400.

Hartman CM & Banks HP. 1980. Pitting in *Psilophyton dawsonii*, an early Devonian trimerophyte. Amer. J. Bot. 67: 400–412.

Hilton J & Bateman RM. 2006. Pteridosperms are the backbone of seed-plant phylogeny. J. Torrey Bot. Soc. 133: 119–168.

Hoffman LA & Tomescu AMF. 2013. An early origin of secondary growth: *Franhueberia gerriennei* gen. et sp. nov. from the Lower Devonian of Gaspé (Quebec, Canada). Amer. J. Bot. 100: 754–763.

Jones TP & Rowe NP. 1999. Fossil plants and spores: modern techniques. The Geological Society, London. 396 pp.

Kenrick P. 1999. Opaque petrifaction techniques. In: Jones TP & Rowe NP (eds.), Fossil plants and spores: modern techniques: 8–91. The Geological Society, London.

Kenrick P & Crane PR 1991. Water-conducting cells in early fossil land plants: implications for the early evolution of tracheophytes. Bot. Gaz. 152: 335–356.

Kenrick P & Crane PR. 1997a. The origin and early diversification of land plants: a cladistic study. Smithsonian Series in Comparative Evolutionary Biology. Smithsonian Institution Press, Washington.

Kenrick P & Crane PR. 1997b. The origin and early evolution of plants on land. Nature 389: 33–39.

Kenrick P & Edwards D. 1988. The anatomy of Lower Devonian *Gosslingia breconensis* Heard based on pyritized axes, with some comments on the permineralization process. Bot. J. Linn. Soc. 97: 95–123.

Kenrick P, Edwards D & Dales RC. 1991. Novel ultrastructure in water-conducting cells of the Lower Devonian plant *Sennicaulis hippocrepiformis*. Palaeontology 34: 751–766.

Kenrick P, Wellman CH, Schneider H & Edgecombe GD. 2012. A timeline for terrestrialization: consequences for the carbon cycle in the Palaeozoic. Phil. Trans. Roy. Soc. London B367: 519–536.

Labandeira CC. 2005. Invasion of the continents: cyanobacterial crusts to tree-inhabiting arthropods. Trends Ecol. Evol. 20: 253–262.

Lak M, Néraudeau D, Nel A, Cloetens P, Perrichot V, Tafforeau P. 2008. Phase contrast X-ray synchrotron imaging: Opening access to fossil inclusions in opaque amber. Microsc. Microanal. 14: 251–259.

Li CS. 1990. *Minarodendron cathaysiense* (gen. et comb. nov.), a lycopod from the late Middle Devonian of Yunnan, China. Palaeontograph. B220: 97–117.

Ligrone R & Duckett JG. 1996. Development of water-conducting cells in the antipodal liverwort *Symphyogyna brasiliensis* (Metzgeriales). New Phytol. 132: 603–615.

Lucas WJ, Groover A, Lichtenberger R, Furuta K, Yadav S-R, Helariutta Y, He X-Q, Fukuda H, Kang J, Brady SM, Patrick JW, Sperry J, Yoshida A, Ana-Flor Lopez-Millàn A-F, Grusak MA & Kachroo P. 2013 The plant vascular system: evolution, development and functions. J. Integr. Plant Biol. 55: 294–388.

Meyer-Berthaud B, Soria A & Decombeix AL. 2010. The land plant cover in the Devonian: a reassessment of the evolution of the tree habit. In: Vecoli M, Clément G & Meyer-Berthaud B (eds.), The terrestrialization process: modelling complex interactions at the biosphere-geosphere interface: 5–70. The Geological Society, London.

Mosbrugger V. 1990. The tree habit in land plants. Lecture Notes in Earth Science 28: 1–161.

Niklas KJ. 1985. The evolution of tracheid diameter in early vascular plants and its implications on the hydraulic conductance of the primary xylem strand. Evolution 39: 1110–1122.

Niklas KJ. 1997. The evolutionary biology of plants. University of Chicago Press, Chicago.

Pittermann J, Sperry JS, Hacke UG, Wheeler JK & Sikkema EH. 2006. Inter-tracheid pitting and the hydraulic efficiency of conifer wood: the role of tracheid allometry and cavitation protection. Amer. J. Bot. 93: 1265–1273.

Pittermann J. 2010. The evolution of water transport in plants: an integrated approach. Geobiology 8: 112–139.

Rasband WS. 2012. ImageJ. U.S. National Institutes of Health, Bethesda, Maryland, USA. http://imagej.nih.gov/ij/. 1997–2012.

Rothwell GW, Sanders H, Wyatt SE & Lev-Yadun S. 2008. A fossil record for growth regulation: The role of auxin in wood evolution. Ann. Missouri Bot. Gard. 95: 121–134.

Sperry JS. 2003. Evolution of water transport and xylem structure. Int. J. Plant Sci. 164: S115–S127.

Spicer R & Groover A. 2010. Evolution of development of vascular cambia and secondary growth. New Phytol. 186: 577–592.

Stein WE. 1993. Modeling the evolution of stelar architecture in vascular plants. Int. J. Plant Sci. 154: 229–263.

Stein WE, Berry CM, Hernick LV & Mannolini F. 2012. Surprisingly complex community discovered in the mid-Devonian fossil forest at Gilboa. Nature 483: 78–81.

Stein WE, Mannolini F, Hernick LV, Landing E & Berry CM. 2007. Giant cladoxylopsid trees resolve the enigma of the Earth's earliest forest stumps at Gilboa. Nature 446 (7138): 904–907.

Strullu-Derrien C, Ducassou C, Ballèvre M, Dabard MP, Gerrienne P, Lardeux H, Le Hérissé A, Robin C, Steemans P & Strullu DG. 2010. The early land plants from the Armorican Massif: sedimentological and palynological considerations on age and environment. Geol. Mag. 147: 830–843.

Strullu-Derrien C, Kenrick P, Tafforeau P, Badel E, Cochard H, Bonnemain JL, Le Hérissé A & Lardeux H. (submitted). The earliest wood and its hydraulic properties documented in 407 million-year-old fossils using synchrotron microtomography.

Tafforeau P & Smith TM. 2008. Nondestructive imaging of hominoid dental microstructure using phase contrast X-ray synchrotron microtomography. J. Human Evol. 54: 272–278.

Taylor TN, Taylor EL & Krings M. 2009. Paleobotany: the biology and evolution of fossil plants. Ed. 2. Elsevier, Academic Press, New York.

Tyree MT & Zimmermann MH. 2002. Xylem structure and the ascent of sap. Springer, Berlin.

Wang DM, Hao SG & Wang Q. 2003. Tracheid ultrastructure of *Hsua deflexa* from the Lower Devonian Xujiachong Formation of Yunnan, China. Int. J. Plant Sci. 164: 415–427.

Wilson JP & Fischer WW. 2011. Hydraulics of *Asteroxylon mackei* [*sic*], an early Devonian vascular plant, and the early evolution of water transport tissue in terrestrial plants. Geobiol. 9: 121–130.

Wilson JP & Knoll AH. 2010. A physiologically explicit morphospace for tracheid-based water transport in modern and extinct seed plants. Paleobiol. 36: 335–355.

Wilson JP, Knoll AH, Holbrook NM & Marshall CR. 2008. Modeling fluid flow in *Medullosa*, an anatomically unusual Carboniferous seed plant. Paleobiol. 34: 472–493.

Accepted: 8 September 2013

IAWA Journal 34 (4), 2013: 352–364

AXIAL CONDUIT WIDENING IN WOODY SPECIES:
A STILL NEGLECTED ANATOMICAL PATTERN

Tommaso Anfodillo*, Giai Petit and **Alan Crivellaro**

Università degli Studi di Padova, Dipartimento Territorio e Sistemi Agro Forestali,
Viale dell'Università 16, 35020 Legnaro (PD), Italy
*Corresponding author; e-mail: tommaso.anfodillo@unipd.it

ABSTRACT

Within a tree the lumen of the xylem conduits varies widely (by at least 1 order of magnitude). Transversally in the stem conduits are smaller close to the pith and larger in the outermost rings. Axially (*i.e.* from petioles to roots) conduits widen from the stem apex downwards in the same tree ring. This axial variation is proposed as being the most efficient anatomical adjustment for stabilizing hydraulic path-length resistance with the progressive growth in height. The hydrodynamic (*i.e.* physical) constraint shapes the whole xylem conduits column in a very similar way in different species and environments. Our aim is to provide experimental evidence that the axial conduit widening is an ineluctable feature of the vascular system in plants. If evolution has favoured efficient distribution networks (*i.e.* total resistance is tree-size independent) the axial conduit widening can be predicted downwards along the stem. Indeed, in order to compensate for the increase in path length with growth in height the conduit size should scale as a power function of tree height with an exponent higher than 0.2. Similarly, this approach could be applied in branches and roots but due to the different lengths of the path roots-leaves the patterns of axial variations of conduit size might slightly deviate from the general widening trend. Finally, we emphasize the importance of sampling standardization with respect to tree height for correctly comparing the anatomical characteristics of different individuals.

Keywords: Tree height, hydraulic resistance, xylem, evolution.

INTRODUCTION

The identification and classification of wood anatomical traits are of fundamental interest to wood anatomists, botanists and plant ecologists. Indeed, variations of such traits in different species and environments largely determine the mechanical and technological properties of wood and thus its economic value. Wood anatomists have therefore established a very detailed set of anatomical traits for describing the astonishing variety of xylem anatomical features that have been evolved by terrestrial plants (Wheeler *et al.* 1989). These anatomical variations form a basis to hypothesize adaptive strategies as drivers of much of the wood anatomical diversity that has resulted through evolution (Carlquist 1975; Baas *et al.* 2004).

© International Association of Wood Anatomists, 2013
Published by Koninklijke Brill NV, Leiden

DOI 10.1163/22941932-00000030

Furthermore, some anatomical traits were used as predictors of the capacity of a given plant to survive in specific environmental conditions. For example, the diameter of the conduits in the xylem is believed to be of major adaptive importance (Sperry *et al.* 2006). In a seminal manuscript (actually the most cited paper ever published in IAWA Journal) (Tyree *et al.* 1994) it is stated that a general trend in xylem conduit diameter can be derived from several anatomical observations (Carlquist 1975): "wet-warm environments tend to favour species with wide conduits whereas cold or dry environments tend to favour species with narrow conduits." A more recent meta-analysis (Wheeler *et al.* 2007) further supported the idea that few wide vessels are associated with tropical environments and many narrow vessels are associated with high latitudes and environments with prolonged periods of low water availability. Interestingly they also noted that vessel diameter is strongly related to habit (*i.e.* height). Shrub species have the highest proportion of narrow vessels (<50 μm) whereas wide vessels (>200 μm) are virtually absent. In trees, on the contrary, wide vessels are very common.

Since the diameter of xylem conduits seems to be a crucial parameter in plant physiological ecology, it is very important to identify why and how conduits size changes within a plant, in plants of different height and in different growing conditions.

Therefore, a wood anatomist/ecologist aiming to study the variation of conduit diameter should ask him-/herself the following questions: where should I localize the sampling point? Does conduit diameter vary in the different parts of plants (roots, branches, stem)? If yes, how large is the variation? How can I standardize the measurements?

We suspect that many readers (if not all) might comment that these questions have already been largely clarified (see for example Tyree & Zimmermann 2002). Notable examples can be found in Sanio (1872) who described "Sanio's trends" in an individual of *Pinus sylvestris*. In that paper the author reported an exhaustive (although not always perfectly clear) description of the trends in tracheids diameter along the stem, branches and roots: in the stem conduits widen basipetally (*i.e.*, increase in diameter axially towards the roots), whereas in roots they are generally wider than in the stem. His observations were very useful and have been substantially confirmed by successive measurements (Zimmermann 1978; Tyree & Zimmermann 2002). In addition, the latter authors found that the axial variation (along the stem) is far from being constant (*i.e.*, linear with the axis length): conduit diameter largely changes near the treetop and then the rate of diameter variation progressively declines becoming rather constant further down at the stem base. This pattern seems to be particularly useful in terms of both hydraulic safety and efficiency, because on the one hand it provides the distal regions of the xylem pathways (where tensions are higher) with the conduits most resistant to cavitation (Hacke *et al.* 2001), on the other it confines the greater part of total hydraulic resistance towards the downstream ends of the flow path (*i.e.* the leaves) (Becker *et al.* 2000; Petit & Anfodillo 2009).

A further and decisive improvement in understanding the structure of the xylem vascular system (and its basic physical requirements) was provided by West *et al.* (1999) with the so-called "WBE model": they were the first to propose an explanation for *why* and *how much* the root-to-leaves column of vascular conduits should vary in width axially, as previously observed. The theoretical approach of West *et al*. (1999) played a pivotal role in understanding axial variation of the pipelines' width. Comments on the model structure, allometric consequences and also the relevant flaws can be easily found elsewhere (Mencuccini 2002; Kozlowski & Konarzewski 2004; McCulloh & Sperry 2005; Coomes 2006; Etienne *et al*. 2006; Petit & Anfodillo 2009).

The WBE model simply proposes a plant formed by a bundle of tubes of the same length running in parallel from the roots to the leaves. Notably the tubes are not cylindrical (as in previous hydraulic models) but are tapered (*i.e*. they widen towards the stem base).

We focus our attention on the straightforward and revealing structure of the WBE model because, differently from all other previously-cited anatomical studies, it allows us to *predict* the variation of conduit diameters along the stem axis or in different individuals throughout ontogenesis.

Our aim is to present the "anatomical structure" of the plant modelled by WBE in detail and to prove that such a predictable pattern is very close to the one observable in nature. We are also convinced that the awareness of the ubiquity of this anatomical feature could help wood anatomists to standardize and functionally compare their measurements.

We are naturally aware that the tree modelled by WBE can only represent an "idealized plant" and that some oversimplifications might be problematic (for example, the assumption that the "tubes" are all of the same length). Our intention is to keep the most valuable idea of the WBE model (and it is really useful) in the knowledge that a lot of work still needs to be done for modelling the hydraulic architecture of trees in detail. Similarly, it is clear that a simple little paper airplane differs enormously from a Boeing 737 but, notably, it does have the same essential property: it flies!

WHY SHOULD CONDUIT DIAMETER VARY WITH PLANT HEIGHT?

Anatomy and physiology are two sides of the same coin. If a general plant requirement must be guaranteed (*e.g*., maintaining leaf efficiency during different life stages) then the anatomy of the plant should be adjusted accordingly. Useful information about anatomical changes can therefore be drawn from models aimed at describing how trees work.

A brief "foray" into the WBE model is needed to explain why conduit diameter should vary with plant height.

In short, the idealized plant modelled by the WBE model is very similar to that proposed by Shinozaki *et al*. (1964): the tree can be seen as a "set of bundles" running in parallel from roots to leaves. These bundles (*i.e*. axial chains of xylem conduits) are connected to a fixed set of leaves (one-to-one in the simplest case). Notably these bundles are believed to be all of the same length: this simplification, which is evidently not true in real plants, will be discussed later.

Overall, the model predicts a complete independence among the different pathways. This condition might be in agreement with the idea of "plant segmentation" (Tyree & Zimmermann 2002) and would bring a selective advantage in ensuring both the

Figure 1. Variation of the earlywood tracheid diameter along the stem axis in the same annual ring (*Picea abies* (L.) Karst.): this variation is referred to as "conduit widening" (direction towards the stem base) or "conduit tapering" if the direction of the water flow is considered. All the sections have the same degree of magnification (indicated at the bottom of the figure: bar = 50 μm).

optimum supply to parts of the plant that are, at the same time, subjected to different metabolic rates (*e.g.* sun or shade leaves) and for better confining system failures (*e.g.* embolisms, pathogens).

In order to maintain the leaf efficiency as constant as possible during ontogeny (*i.e.*, when the tree grows) a new "ingredient" must be introduced to the simple "pipe model": the pipes are not cylindrical but they have a different diameter along the longitudinal axis (*i.e.* they widen basipetally) (Fig. 1). The variation of conduit diameter downwards in the stem (conduit widening, also known as conduit tapering) is predicted to be a power law according to which the variation of conduit diameter (*Dh*) with the distance from the tree top (*L*) (note that it is the inverse of tree height) will be:

$$Dh \propto L^b,$$

where *b* is the exponent of the power function, which accounts for the relative variation. In the case of $b = 0$ there is no axial variation and the shape of the tube is cylindrical (like the original version of the pipe model). The rate of increase in hydraulic resistance with tree height strongly depends on how conduits vary in size along the stem. When xylem cells increase in diameter from the stem apex to the base a higher proportion of resistance is confined towards the apex. The higher the degree of widening, the greater is the magnitude of resistance confined to the apex. This implies that with further stem elongation, widening conduits towards the stem base would allow for a compensatory effect on the path length resistance, the efficiency of which is higher for higher degrees of widening (Becker *et al.* 2000; Petit & Anfodillo 2009) (Fig. 2). Notably, the degree of widening predicted by the WBE model ($b = 0.25$) can be considered as a threshold value above which the independence of the hydraulic resistance from the total path length (*i.e.* tree height) is guaranteed. The WBE theory predicts that evolution has acted in such a way that the degree of widening is the *minimum* required to make the resistance independent of the tree height (Enquist 2003). If true, then the leaf metabolic efficiency will be kept "invariant" through ontogeny, *i.e.*, water supply to the leaves will be similar in both small and very tall trees.

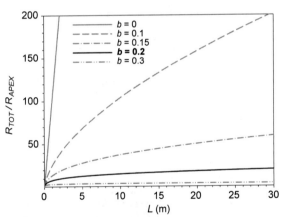

Figure 2. The effect of progressive degree of widening (or tapering) of vascular conduits (*i.e.* the scaling exponent *b* in determining the total hydraulic resistance, relative to the resistance of the most apical conduit (R_{TOT} / R_{APEX}), with the increasing length (L) of a single pipeline. Theory assumes that plants should approach the minimum degree of widening ($b = 0.2$) required to fully compensate for the progressive increase in hydraulic resistance (on the y-axis) with the growth in height (*i.e.* path length on the x-axis). The scaling of the total resistance at different exponent *b* can be predicted on the basis of physical laws.

In order to compare real trees with the "idealized plant" of the WBE model it is necessary to take into account the approximation of the WBE model in estimating the tree height. Doing so the minimum scaling exponent *b*, relative to the scaling of the conduit dimension *versus* the distance from the treetop, assumes the value of about 0.20 (see details in Anfodillo *et al.* 2006). This value (exponent similar to or higher than 0.20) means that in a real tree (not in the idealized plant of the WBE model) the vascular network is structured to compensate for the increase in plant height. Exponents in agreement with the predicted one have been repeatedly measured in plants of different sizes and environments (Anfodillo *et al.* 2006; Coomes *et al.* 2007; Petit *et al.* 2008, 2009; Lintunen & Kalliokoski 2010; Petit *et al.* 2010, 2011; Bettiati *et al.* 2012; Olson & Rosell 2012) (Fig. 3).

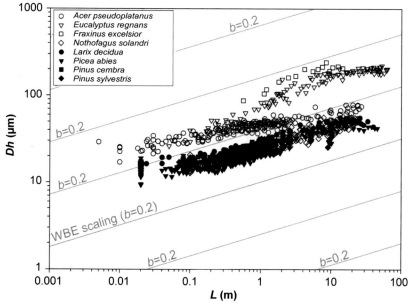

Figure 3. A meta-analysis of the axial variation of xylem conduit diameter (*Dh*) with the distance from the stem apex (*L*) of different species (inset). Data refer to measurements in stems and branches. Literature data are: *Acer pseudoplatanus* from Petit *et al.* (2008); *Eucalyptus regnans* from Petit *et al.* (2010); *Fraxinus excelsior* from Anfodillo *et al.* (2006) and Bettiati *et al.* (2012); *Nothofagus solandri* from Coomes *et al.* (2007); *Larix decidua* from Anfodillo *et al.* (2006) and Petit *et al.* (2009); *Picea abies* from Anfodillo *et al.* (2006), Coomes *et al.* (2007), Petit *et al.* (2009) and Petit *et al.* (2011); *Pinus cembra* from Petit *et al.* (2009); *Pinus sylvestris* from Coomes *et al.* (2007) and original data. The parallel grey lines indicate the power scaling with exponent b = 0.20. Lines change in relation to different intercepts (*i.e.* conduit size at the plant apex).

With the knowledge gained so far it is logical to expect that the relatively largest conduit dimensions at the stem base should be found in tall trees and, in contrast, small plants will have relatively small conduits: this general pattern in conduit size is therefore the *consequence* of the plant size. A meta-analysis on available data shows that conduits at the stem base are generally wider in taller plants (Fig. 4 and see also Wheeler *et al.* 2007)

Figure 4. Variation of conduit diameter at the stem base (in 4 classes: < 20; 20–50; 50–100; 100–200 μm) in relation to averaged class of plant height (4 classes). Data from Fritz Schweingruber relative to 3339 different angiosperm species mainly from the northern hemisphere (available at: www.wsl.ch/dendro/xylemdb/index.php). The tallest plants have formed the widest conduits.

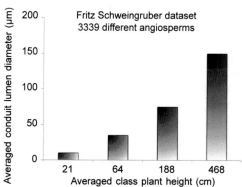

SOLVING THE CHICKEN-EGG DILEMMA

As mentioned above, Tyree *et al.* (1994) wrote that small conduits were generally measured in plants of dry/cold environments and therefore it could be speculated that, in a
low-resources site, natural selection acts against species with relatively large conduits.
However, the previous considerations showed that conduit size is closely correlated to
tree height. Now, since tree height is negatively affected in low-resources sites so plants
are generally small, it is obvious to expect narrower conduits. This circular reasoning
could be solved by disentangling the role of tree size in determining the conduit size
from that of resources' availability.

In a recent elegant analysis Olson and Rosell (2012) compared the conduit dimension
at the stem base, in 142 different species growing in 5 sites within a huge gradient of
water availability (annual precipitation from about 800 to 3500 mm). They then plotted
the conduit dimension of each plant *versus* its stem diameter (which is allometrically
linked to tree height) in each site. In this way they were able to compare the intercept of
each regression line, *i.e.* the average conduit dimension in plants with the *same basal
diameter*. They clearly demonstrated that the conduit size in plants growing in the wettest
and driest sites did not differ. The conclusion is that plants of the drier sites are generally smaller and *therefore* have narrower conduits but, *at the same size* their conduits
substantially do not differ from those of the species growing in the wetter sites.

This places full emphasis on how important it is to standardize the collection of
samples in relation to plant height if the aim is to evaluate the adaptive consequences
of given anatomical structures. The awareness of the ubiquity of the axial widening
pattern could therefore help in correctly interpreting the anatomical traits.

Even more importantly, results indicate that conduit diameter is strictly related to tree
height but little or not at all to the *age* of the plant. We believe that it is time to revise
the common belief that this wood trait is substantially age-dependent. What occurs in
the cambial zone and the cascade of events leading to a final conduit dimension are primarily related to the distance of the cambial cells from the apex (and, obviously, to environmental constraints). Indeed, it was demonstrated that the number of days in which
the forming cells remain in the expansion phase is directly related to distance from the
tree top (L) (Anfodillo *et al.* 2012). Cambial activity and the cell formation is basically
height- and *not age*-related, as confirmed by Petit *et al.* (2008), who showed that apical
shoots from very tall parent *Acer pseudoplatanus* trees, grafted onto 1-year-old root-
stocks, developed vessels of similar sizes to those of young trees of similar height.

One of the important conclusions of this paper is that, when dealing with variation of
wood traits (*e.g.* density, cell dimension, fraction of latewood etc.) within a functional-
based approach, it would be better to change the traditional idea of "age-dependency"
with the more correct concept of "size-dependency".

PERVASIVENESS OF THE PATTERN

One of the most unrealistic assumptions of the WBE model is that it considers an idealized plant formed by "pipes" of the same length. However, in trees branches have
different length: some of them are very close to the ground and others near the treetop

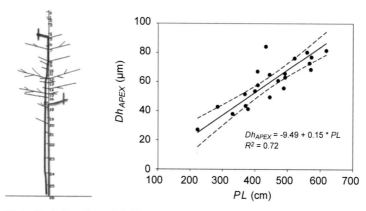

Figure 5. Right: Variation of conduit diameter (*Dh*) at the apex of the branches in relation to the length of the path *PL* (calculated from root to apex). Left: example of two paths (grey thick lines) with different length in a stylized *Fraxinus excelsior* (number of internodes are also indicated). The diameter of apical conduits increases with the increase of the path: this anatomical feature should compensate for the different length of the paths. Shorter paths have smaller apical conduits thus the total hydraulic resistance can be maintained very similarly among different paths (data from Bettiati *et al.* 2012).

(Fig. 5). Thus a basic assumption of the model appeared to be violated in nature. How can a plant cope with the variable length of the paths? The simple underlying idea is that a plant should adjust the structure (*i.e.* anatomy) of the vascular conduits to achieve a condition of equiresistance of all roots-to-leaves paths. If this condition was not satisfied then water would flow mainly throughout the paths with lower hydraulic resistance. But this would be detrimental to achieving similar water supply to all leaves.

There is not much information about the hydraulic architecture of the *whole* tree carried out with appropriate sampling (*i.e.* taking into account the distance of the samples from the tree top). Some examples can be derived from the book by Tyree & Zimmermann (2002) but in many cases the exact sampling distance is not specified thus making interpretation of results difficult. Recent anatomical analyses (Bettiati *et al.* 2012) showed that diameter of apical conduits in branches (collected just below the apex) are significantly different, with the widest diameters found in the longest root-to-leaves paths (Fig. 5). For example, in very short branches in the basal part of the crown (short path) apical conduits are relatively small (2–3 times smaller than the conduits in the longest paths/branches). This seems a very simple and coherent adjustment in order to guarantee similar resistance in all paths and therefore equal water delivery to the different parts of the crown.

Generally, it clearly emerges that the anatomical feature of one part of a plant (*e.g.*, an apex of a certain branch) is strictly linked to the anatomy of the *whole* individual. Thus in spite of the near independence of all different paths they are anatomically structured in order to supply all the leaves in the different parts of the crown at similar rates: this result is equivalent to that obtained considering all the tubes of the same length (as in the simple WBE idealized plant).

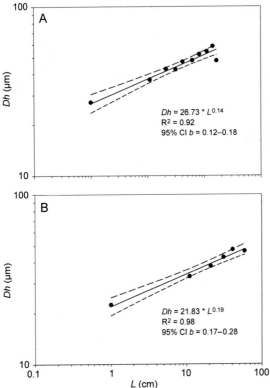

Figure 6. Variation of conduit diameter (*Dh*) with the distance from the apex in two different branches of a 9.5 m tall poplar tree (*Populus × canadensis*). The exponent *b* slightly changes (the 95% CI is also indicated). In less than 2.0 m in length the conduit diameter changes by a factor of 2.5 (from about 20 to 50 μm).

Since the "tubes" run from roots to leaves, the same general pattern of conduit widening should in principle be preserved along the longitudinal axes of both branches and roots. Systematic measurements of conduit widening in branches are even more difficult to find than the axial variation of conduit size in stems.

However, measurements of hydraulic permeability (a proxy for conduit diameter) in branches generally showed an increase with branch diameter (*i.e.* towards the stem) (Tyree & Alexander 1993; Jerez *et al.* 2004; Sellin & Kupper 2007) thus demonstrating that conduits become wider towards the branch base.

An analysis on the axial variation of conduit diameter in branches of a poplar tree showed a common pattern of widening compatible with the value of 0.20 but short branches also showed a lower degree of tapering (0.16–0.14) (Fig. 6).

Similarly to stem anatomy the variation of conduit dimension along the branch strictly depends on the distance from the branch tip. The conduit diameter might easily vary by a factor of about 2 for a variation of 1 m in position of the sample towards the stem.

It is evident how important it is to also consider the pattern of conduit widening in branches. Any measurement, for example, of hydraulic conductivity, which is taken on small pieces of branches, must be normalized for the distance from the branch tip. Otherwise the results might be dependent only on the position of the sample and thus become meaningless.

Xylem conduits are believed to continue to increase in diameter downwards also in roots. Indeed, in general, vessel diameter and length in woody roots exceed those in stems of comparable diameter (Tyree & Zimmermann 2002), thus the widest cells of the whole plant are found in roots. This pattern was observed in shrubs of cold deserts where the mean vessel diameter was about 2 times wider than in the stem (Kolb & Sperry 1999), in nine Mediterranean woody species (Martinez-Vilalta *et al.* 2002) where the diameter of conduits was always wider in roots than in stems, and in conifers (Petit *et al.* 2009).

However, in very short superficial roots it is not uncommon to find relatively narrow cells (even narrower than in the stem). This is probably due to the fact that in roots (as occurs in branches) the length of the paths might differ significantly so the plant would compensate for possible differing hydraulic resistance by adjusting the conduit dimension accordingly (*i.e.* narrower conduits in shorter roots).

Notably in the roots the allometric relationships of root diameter *versus* distance from the tree top has an opposite sign compared to the stem (in roots the exponent is negative) showing that the mechanical constraints in roots differ from those of the stem and branches. Nonetheless the hydraulic constraints in roots are similar to the other organs so the degree of widening is supposed to be similar in all parts of the plants. Measurements in roots indicate that conduit widening seems to be a stable property of the whole xylem architecture, with the widest conduits very close to the root tips (as predictable from the hydraulic requirements) (Petit *et al.* 2009). However, further measurements are needed to clarify the pattern of conduit size in roots because they are characterized by a branching network with huge variations in the length of the different conductive paths.

A STILL UNANSWERED QUESTION

One of the most intriguing physiological questions related to axial widening is: how can a plant so precisely regulate the axial conduit dimensions along a path that may exceed 150 m in length (stem and roots in the tallest trees)?

Wood formation in trees is a dynamic process, strongly affected by environmental conditions, including nutrient availability and climatic changes (Oribe *et al.* 2003; Arend & Fromm 2007; Sorce *et al.* 2013; Prislan *et al.* 2013). Despite the relevant role of cambium tissues in plants, few studies have dealt with the molecular and structural mecanisms at the basis of its functionality (Deslauriers *et al.* 2009; Berta *et al.* 2010). According to a recent study, wider cells along the stem are those staying longer in the distension phase during xylogenesis (Anfodillo *et al.* 2012). Therefore, it is likely that plants adopt a mechanism to modulate the time for cell enlargement to precisely design xylem conduits optimally widened from the stem apex downwards for hydraulic purposes.

A candidate hypothesis proposes the polar transportation of phyto-hormones, particularly auxin (IAA), as the mechanism of control of the dimension of xylem cells along the stem (Aloni 2001; Aloni *et al.* 2003). The IAA is produced mainly in the developing buds and shoots (Uggla *et al.* 1998; Scarpella & Meijer 2004) and is transported basipetally along the cambial zone (Sundberg *et al.* 2000) from leaves to

roots (Aloni 2001; Muday & DeLong 2001). Moreover, the IAA concentration generally decreases from the stem apex to the base (Lovisolo *et al*. 2002). According to the
six point hypothesis (Aloni & Zimmermann 1983), high concentrations of IAA would
accelerate the cellular differentiation, thus reducing the time period for the cellular
distension phase.

This new research topic should be promoted given the pivotal role in regulating
the efficiency of water transport. We hope that this simple manuscript might encour-
age some scientists to clarify the physiological mechanisms related to axial conduit
widening.

CONCLUSIONS

The xylem conduit size in stems, branches and roots appeared invariably dependent
on the distance from the top: moving downwards (basipetally) the conduits are gradually wider and this is a necessary anatomical feature for stabilizing the hydrodynamic
resistance with tree height.

Thus the requirement for maintaining the efficiency of water transport throughout
the plant ontogenesis in all crown parts is achieved by shaping a widened anatomical
structure. We believe that the axial conduit widening can no longer be neglected because
it offers a clear and universal physiological explanation of the anatomical changes
carefully observed by Sanio more than a century ago.

ACKNOWLEDGEMENTS

The manuscript was funded by the University of Padova, project UNIFORALL (CPDA110234). It was
also inspired and supported by the EU COST Action FP1106. Alan Crivellaro was supported by the
University of Padova (Assegno di Ricerca Junior CPDR124554/12). The authors thank A.Garside for
checking the English text.

REFERENCES

Aloni R. 2001. Foliar and axial aspects of vascular differentiation: hypotheses and evidence.
    J. Plant Growth Regul. 20: 22–34.
Aloni R, Schwalm K, Langhans M & Ullrich C. 2003. Gradual shifts in sites of free-auxin pro-
    duction during leaf-primordium development and their role in vascular differentiation and
    leaf morphogenesis in *Arabidopsis*. Planta 216: 841–853.
Aloni R & Zimmermann MH. 1983. The control of vessel size and density along the plant axis:
    a new hypothesis. Differentiation 24: 203–208.
Anfodillo T, Carraro V, Carrer M, Fior C & Rossi S. 2006. Convergent tapering of xylem conduits
    in different woody species. New Phytol. 169: 279–290.
Anfodillo T, Deslauriers A, Menardi R, Tedoldi L, Petit G & Rossi S. 2012. Widening of xylem
    conduits in a conifer tree depends on the longer time of cell expansion downwards along
    the stem. J. Exp. Bot. 63: 837–845.
Arend M & Fromm J. 2007. Seasonal change in the drought response of wood cell development
    in poplar. Tree Physiol. 27: 985–992.
Baas P, Ewers FW, Davies SD & Wheeler EA. 2004. The evolution of xylem physiology.
    In: Hemsley AR & Poole I (eds.), Evolution of plant physiology from whole plants to ecosys-
    tems: 273–296. Linnean Society Symposium Series 21. London, Elsevier Academic Press.

Becker P, Gribben RJ & Lim CM. 2000. Tapered conduits can buffer hydraulic conductance from path-length effects. Tree Physiol. 20: 965–967.

Berta M, Giovannelli A, Sebastiani F, Camussi A & Racchi ML. 2010. Transcriptome changes in the cambial region of poplar (*Populus alba* L.) in response to water deficit. Plant Biol. 12: 341–354.

Bettiati D, Petit G & Anfodillo T. 2012. Testing the equi-resistance principle of the xylem transport system in a small ash tree: empirical support from anatomical analyses. Tree Physiol. 32: 171–177.

Carlquist S. 1975. Ecological strategies of xylem evolution. University of California Press, Berkeley.

Coomes DA. 2006. Challenges to the generality of WBE theory. Trends Ecol. Evol. 21: 593–596.

Coomes DA, Jenkins KL & Cole LES. 2007. Scaling of tree vascular transport systems along gradients of nutrient supply and altitude. Biol. Lett. 3: 86–89.

Deslauriers A, Giovannelli A, Rossi S, Castro G, Fragnelli G & Traversi L. 2009. Intra-annual cambial activity and carbon availability in stem of poplar. Tree Physiol. 29: 1223–1235.

Enquist BJ. 2003. Cope's Rule and the evolution of long-distance transport in vascular plants: allometric scaling, biomass partitioning and optimization. Plant Cell Environm. 26: 151–161.

Etienne RS, Apol MEF & Olff HAN. 2006. Demystifying the West, Brown & Enquist model of the allometry of metabolism. Funct. Ecol. 20: 394–399.

Hacke UG, Sperry JS, Pockman WT, Davis SD & McCulloch KA. 2001. Trends in wood density and structure are linked to prevention of xylem implosion by negative pressure. Oecologia 126: 457–461.

Jerez M, Dean T, Roberts S & Evans D. 2004. Patterns of branch permeability with crown depth among loblolly pine families differing in growth rate and crown size. Trees 18: 145–150.

Kolb KJ & Sperry JS. 1999. Transport constraints on water use by the Great Basin shrub, *Artemisia tridentata*. Plant Cell Environm. 22: 925–935.

Kozlowski J & Konarzewski M. 2004. Is West, Brown and Enquist's model of allometric scaling mathematically correct and biologically relevant? Functional Ecology 18: 283–289.

Lintunen A & Kalliokoski T. 2010. The effect of tree architecture on conduit diameter and frequency from small distal roots to branch tips in *Betula pendula*, *Picea abies* and *Pinus sylvestris*. Tree Physiol. 30: 1433–1447.

Lovisolo C, Hartung W & Schubert A. 2002. Whole-plant hydraulic conductance and root-to-shoot flow of abscisic acid are independently affected by water stress in grapevines. Functional Plant Biology 29: 1349–1356.

Martinez-Vilalta J, Prat E, Oliveras L & Pinol J. 2002. Xylem hydraulic properties of roots and stems of nine Mediterranean woody species. Oecologia 133: 19–29.

McCulloh KA & Sperry JS. 2005. Patterns in hydraulic architecture and their implications for transport efficiency. Tree Physiol. 25: 257–267.

Mencuccini M. 2002. Hydraulic constraints in the functional scaling of trees. Tree Physiol. 22: 553–565.

Muday GK & DeLong A. 2001. Polar auxin transport: controlling where and how much. Trends in Plant Science 6: 535–542.

Olson ME & Rosell JA. 2012. Vessel diameter–stem diameter scaling across woody angiosperms and the ecological causes of xylem vessel diameter variation. New Phytol. 197: 1204–1213.

Oribe Y, Funada R & Kubo T. 2003. Relationships between cambial activity, cell differentiation and the localization of starch in storage tissues around the cambium in locally heated stems of *Abies sachalinensis* (Schmidt) Masters. Trees 17: 185–192.

Petit G & Anfodillo T. 2009. Plant physiology in theory and practice: An analysis of the WBE model for vascular plants. J. Theor. Biol. 259: 1–4.

Petit G, Anfodillo T, Carraro V, Grani F & Carrer M. 2011. Hydraulic constraints limit height growth in trees at high altitude. New Phytol. 189: 241–252.

Petit G, Anfodillo T & De Zan C. 2009. Degree of tapering of xylem conduits in stems and roots of small *Pinus cembra* and *Larix decidua* trees. Botany 87: 501–508.

Petit G, Anfodillo T & Mencuccini M. 2008. Tapering of xylem conduits and hydraulic limitations in sycamore (*Acer pseudoplatanus*) trees. New Phytol. 177: 653–664.

Petit G, Pfautsch S, Anfodillo T & Adams MA. 2010. The challenge of tree height in *Eucalyptus regnans*: when xylem tapering overcomes hydraulic resistance. New Phytol. 187: 1146–1153.

Prislan P, Čufar K, Koch G, Schmitt U & Gričar J. 2013. Review of cellular and subcellular changes in cambium. IAWA J. 34: 391–407.

Sanio K. 1872. Über die Größe der Holzzellen bei der gemeinen Kiefer (*Pinus sylvestris*). J. Wiss. Bot. 8: 401–420.

Scarpella E & Meijer AH. 2004. Pattern formation in the vascular system of monocot and dicot plant species. New Phytol. 164: 209–242.

Sellin A & Kupper P. 2007. Effects of enhanced hydraulic supply for foliage on stomatal responses in little-leaf linden (*Tilia cordata* Mill.). Eur. J. Forest Res. 126: 241–251.

Shinozaki K, Yoda K, Hozumi K & Kira T. 1964. A quantitative analysis of plant form - The pipe model theory I. Basic analyses. Jap. J. Ecol. 14: 94–105.

Sorce C, Giovannelli A, Sebastiani L & Anfodillo T. 2013. Hormonal signals involved in the regulation of cambial activity, xylogenesis and vessel patterning in trees. Plant Cell Rep. 32: 885–898.

Sperry JS, Hacke UG & Pittermann J. 2006. Size and function in conifer tracheids and angiosperm vessels. Amer. J. Bot. 93: 1490–1500.

Sundberg B, Uggla C & Tuominen H. 2000. Cambial growth and auxin gradients. In: Savidge RA, Barnett JR & Napier R (eds.), Cell and molecular biology of wood formation: 169–188. BIOS Scientific Publishers Ltd., Oxford.

Tyree MT & Alexander J. 1993. Hydraulic conductivity of branch junctions in three temperate tree species. Trees 7: 156–159.

Tyree MT, Davis SD & Cochard H. 1994. Biophysical perspectives of xylem evolution: is there a tradeoff of hydraulic efficiency for vulnerability to dysfunction? IAWA J. 15: 335–360.

Tyree MT & Zimmermann MH. 2002. Xylem structure and the ascent of sap. Springer, Berlin.

Uggla C, Mellerowicz EJ & Sundberg B. 1998. Indole-3-acetic acid controls cambial growth in Scots pine by positional signaling. Plant Physiol. 117: 113–121.

West GB, Brown JH & Enquist BJ. 1999. A general model for the structure and allometry of plant vascular systems. Nature 400: 664–667.

Wheeler EA, Baas P & Gasson PE (eds.). 1989. IAWA List of microscopic features for hardwood identification. IAWA Bull. n.s. 10: 219–332.

Wheeler EA, Baas P & Rodgers S. 2007. Variations in dicot wood anatomy: a global analysis based on the InsideWood database e.a. IAWA J. 28: 229–258.

Zimmermann MH. 1978. Hydraulic architecture of some diffuse-porous trees. Can. J. Bot. 56: 2286–2295.

Accepted: 29 August 2013

IAWA Journal 34 (4), 2013: 365–390

BRILL

HYDRAULIC AND BIOMECHANICAL OPTIMIZATION IN NORWAY SPRUCE TRUNKWOOD – A REVIEW

Sabine Rosner

Institute of Botany, BOKU Vienna, Gregor Mendel Str. 33, 1180-Vienna, Austria
E-mail: sabine.rosner@boku.ac.at

ABSTRACT

Secondary xylem (wood) fulfills many of the functions required for tree survival, such as transport of water and nutrients, storage of water and assimilates, and mechanical support. The evolutionary process has optimized tree structure to maximize survival of the species, but has not necessarily optimized the wood properties needed for lumber. Under the impact of global warming, knowledge about structure-function relationships in tree trunks will become more and more important in order to prognosticate survival prospects of a species, individuals or provenances. Increasing our knowledge on functional wood anatomy can also provide valuable input for the development of reliable, fast, and at best quasi-non-destructive (*e.g.* wood coring of mature trunks) indirect screening techniques for drought susceptibility of woody species. This review gives an interdisciplinary update of our present knowledge on hydraulic and biome-chanical determinants of wood structure within and among trunks of Norway spruce (*Picea abies* (L.) Karst.), which is one of Europe's economically most important forest tree species. It summarizes what we know so far on 1) within-ring variability of hydraulic and mechanical properties, 2) structure-function relationships in mature wood, 3) mechanical and hydraulic demands and their tradeoffs along tree trunks, and 4) the quite complex wood structure of the young trunk associated with mechanical demands of a small tree. Due to its interdisciplinary nature this review is addressed to physiologists, foresters, tree breeders and wood technologists.

Keywords: Biomechanics, functional anatomy, hydraulic vulnerability, Norway spruce, *Picea abies*, tracheid, water transport, wood density.

INTRODUCTION

Global change is expected to increase the frequency of extreme weather and climate events such as heat waves, drought periods and storms (Schär *et al.* 2004; Salinger 2005; IPCC 2012; Semenov 2012). As a consequence, plant productivity and growth will be affected in many regions (Ciais *et al.* 2005; Bréda *et al.* 2006; Reich & Oleksyn 2008; Kullman & Öberg 2009; Choat *et al.* 2012; Williams *et al.* 2012). Continued refinement and extension of models linking between plant hydraulics, mechanics and ecosystem functioning will help to gain information on broader patterns in productivity that are

© International Association of Wood Anatomists, 2013
Published by Koninklijke Brill NV, Leiden

DOI 10.1163/22941932-00000031

related to plant water use (Geßler *et al.* 2007; McDowell *et al.* 2008; Sperry *et al.* 2008; Meinzer *et al.* 2009, 2010; Barnard *et al.* 2011). This knowledge can be applied to prognosticate survival prospects of woody species (Pockman & Sperry 2000; Maherali *et al.* 2004; Hukin *et al.* 2005; Dalla-Salda *et al.* 2009, 2011; Choat *et al.* 2012), or to screen for less drought-sensitive clones (Rozenberg *et al.* 2002; Monclus *et al.* 2005; Cochard *et al.* 2007; Rosner *et al.* 2007, 2008), or provenances (Peuke *et al.* 2002; Kapeller *et al.* 2012). Screening for hydraulic and mechanical performance demands, however, easily applicable methods. Increasing our knowledge on structure-function relationships in trunkwood can provide valuable input for the search on reliable, fast, and at best quasi-non-destructive (*e.g.* wood coring of mature trunks) methodologies for indirect assessment of wood biological functions. This review gives an overview on our present knowledge on structure-function relationships in the trunkwood of Norway spruce (*Picea abies* (L.) Karst.), which is one of central and northern Europe's economically most important forest tree species (Bergh *et al.* 2005). Norway spruce can be found up to the alpine timberline (Mayr *et al.* 2002, 2003) as well as in the far north (Tollefsrud *et al.* 2008). Problems associated with weak adaptation of Norway spruce to environmental conditions in lowland regions, such as top dieback, damage due to wind and snow loads and subsequent bark beetle mass outbreaks which are favoured by drought stress of the host trees frequently cause enormous economic losses (Solberg 2004; Andreassen *et al.* 2006; Schlyter *et al.* 2006).

Interdisciplinary research on structure-function relationships within conifer trunks started in the late 90-ties when Mencuccini *et al.* (1997) published their work on biomechanical and hydraulic determinants of tree structure in Scots pine (*Pinus sylvestris* L.). Experimental studies on structure-function relationships within conifer trunks are, however, still quite scarce and much more literature exists on pine species than on spruce species (Lachenbruch *et al.* 2011). The most entirely investigated conifer species are Douglas fir (*Pseudotsuga menziesii* (Mirb.) Franco) (*e.g.* Domec & Gartner 2001; Spicer & Gartner 2001; Domec & Gartner 2002a, 2002b; Dunham *et al.* 2007; Domec *et al.* 2009) and Ponderosa pine (*Pinus ponderosa* Dougl. *ex* Laws.) (*e.g.* Bouffier *et al.* 2003; Domec & Gartner 2003; Domec *et al.* 2005; Domec *et al.* 2009; Barnard *et al.* 2011). Jagels *et al.* (2003) and Jagels and Visscher (2006) provided us also insights into the quite unique structure-function relationships within the main trunk of *Metasequoia* (*M. glyptostroboides* Hu *et* Cheng). Mayr *et al.* (2002) were the first to relate hydraulic performance in Norway spruce leader shoots to wood structure and discussed the results upon mechanical demands before Rosner and colleagues published their work series on biomechanical and hydraulic optimization within and among Norway spruce trunks (Rosner *et al.* 2006, 2007, 2008, 2009, 2010; Rosner & Karlsson 2011; Rosner *et al.* 2012).

The dataset developed during the past years on structure-function relationships in Norway spruce trunkwood is quite unique; Norway spruce was used *e.g.* as a model species to develop new analysis methods to assess hydraulic vulnerability and interdisciplinary research combining wood anatomy, wood physiology and wood technology was helpful to get a broader view of structure-function relationships within the tree trunk. The aim of this review is to give an update of our present knowledge on hydraulic and biomechanical determinants of wood structure within and among Norway spruce trunks and is addressed to physiologists, foresters, tree breeders as well as to wood technologists.

The tracheid: a multitasking wood element

Tracheids fulfill many of the functions required for tree survival, such as water transport, water storage and mechanical support. Norway spruce secondary xylem (wood) consists to more than 90 % of tracheids (Fig. 1), the remaining tissue is parenchyma (Brändström 2001). Cell wall layers of tracheids consist of 40–50 % cellulose, 25–30 % lignin and 20–25 % hemicellulose (Plomion *et al.* 2001). Cellulose chains are arranged in microfibrils, which are unordered in the primary wall, but highly ordered in the thickest layer of the secondary cell wall (S_2) (Booker & Sell 1998). Reviews on ultra-structural features of the different cell wall layers with special reference to Norway spruce tracheids can be found elsewhere (Neagu *et al.* 2006; Jungnikl *et al.* 2008; Salmén & Burgert 2009). Each tracheid is connected with its adjacent tracheids by bordered pits (Fig. 2) consisting of a porous membrane held in a pit chamber. Pit membranes in Norway spruce are of the torus-margo type, owing a thin and porous margo with a thickened torus (Liese & Bauch 1967; Greaves 1973; Gregory & Petty 1973).

According to its hydraulic and biomechanical tasks within the trunk, the morphology of a tracheid varies considerably. Variability in anatomical structure and chemical composition of cell walls within a stem includes within-ring differences, known as earlywood and latewood (Fig. 2, 3b & 3c), radial variations resulting

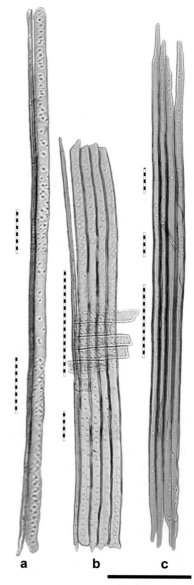

a   b   c

Figure 1. Macerated Norway spruce tracheids of the last latewood cell row and the adjacent first earlywood cell rows formed in the following growing season (**a** & **b**) and tracheids of the earlywood formed later in the growing season (**c**). Macerated tracheid samples came from young Norway spruce trees (field age 5 years) with different growth characteristics; this explains their differences in size. Hatched bars indicate regions of cross-field pitting. The reference bar represents 200 µm. The figure is modified after Rosner *et al.* (2007).

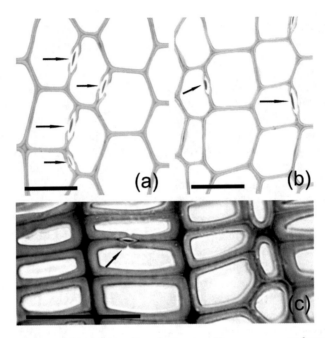

Figure 2. Transverse semi-thin sections (2 μm) of mature Norway spruce earlywood with open (a) and closed (aspirated) bordered pits (b) and of Norway spruce latewood (c) embedded in Technovit 7100 (Heraeus Kulzer GmbH, Wehrheim, Germany) and stained with toluidine blue. Arrows point at the darker stained tori of the pit membranes. Toluidine blue stains the torus dark blue/purple, which gives contrast to the light blue stained pit chamber walls and secondary cell walls of the tracheids (a, b). Bordered pits are merely present at radial cell walls (a, b) but can be as well found in tangential cell walls (c). Reference bars represent 50 μm. Technical details can be found in Rosner *et al.* (2010).

from cambial maturation and sapwood aging, and differences associated with height position within the trunk (Zobel & Van Buijtenen 1989; Gartner 1995; Lundgren 2004; Havimo *et al.* 2008; Lachenbruch *et al.* 2011). A general concept in conifer species such as Norway spruce is that the radial cell wall of a tracheid is thicker than the tangential cell wall (Brändström 2001). Radial width, tangential width and cell-wall thickness vary even along the tracheid length and, moreover, at each contact with rays a sudden change in tangential width and cell wall thickness is observed (Sarén *et al.* 2001; Neagu *et al.* 2006). Bordered pits, which are found more frequently in the radial cell walls, can thus be seen as natural irregularities influencing the biomechanical properties of the tracheids (Sirviö & Kärenlampi 1998).

Water transport in xylem is achieved under negative pressure: general structural concepts

Water is transported from soil to the transpiring leaves under negative hydrostatic pressure (Zimmermann 1983; Richter 2001; Tyree 2003; Domec 2011), requiring high mechanical strength of the tracheid cell walls in order to avoid implosion, and sufficient

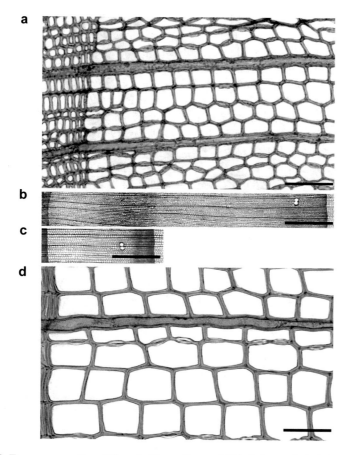

Figure 3. Transverse sections (20 μm) of juvenile wood (1st–2nd annual ring) from the tree top (**a, b**) and mature wood (17th–19th annual ring) from the lower tree trunk (**c, d**) of Norway spruce. Reference bars represent 50 μm (a, d) and 200 μm (b, c).

safety factors against the breakage of the water column (cavitation). According to the air-seeding hypothesis (Tyree & Zimmermann 2002), cavitation due to drought stress will occur when the pressure difference between water in a conduit and surrounding air exceeds the capillary forces at the air-water interface in the conduit wall. Under these conditions, air will be pulled into the conduit and the air bubble will cause a phase change to vapor.

High wood density is supposed to be a common strategy to guarantee low vulner-ability to cavitation. In theory, greater resistance to cavitation requires a safer design for resisting implosion, because the cell walls have to withstand higher tensile strain before cavitation occurs (Hacke *et al.* 2001; Hacke & Sperry 2001). Tensile stresses in a water-filled conduit are supposed to increase with decreasing squared double cell wall (t) to span ratio (b), termed conduit wall reinforcement [$(t/b)^2$], based on the fact that both mechanical strength and stiffness increase with increasing wood density (Hacke *et al.* 2001; Hacke & Sperry 2001; Pitterman *et al.* 2006; Sperry *et al.* 2006; Domec *et al.* 2009). An increase in density, however, goes along with a decrease in hydraulic efficiency, because it is achieved mainly by narrowing lumen diameter rather than increasing cell wall thickness (Hannrup *et al.* 2004; Pitterman *et al.* 2006; Sperry *et al.* 2006). Resistance to drought-induced cavitation is however not directly related to wood density (Hacke *et al.* 2001; Hacke & Sperry 2001; Domec & Gartner 2002a), but is more closely related to the function of the pit membrane properties (Tyree & Sperry 1989; Sperry & Tyree 1990; Tyree & Ewers 1991; Sperry 1995; Sperry & Ikeda 1997; Hacke & Sperry 2001; Jansen *et al.* 2012).

When sapwood from the lower trunk of older trees is compared to sapwood of very young trees (Rosner *et al.* 2007, 2008), the relationship between wood density and vulnerability to cavitation actually exists in Norway spruce: P_{50} (the water potential necessary to cause 50 % loss of hydraulic conductivity) increases with decreasing wood density (Fig. 4). However, does the construction of such a relationship help to select individuals with high drought susceptibility or is it helpful to understand strategies or mechanisms behind hydraulic vulnerability? A closer view is necessary to learn *e.g.* which tracheids are the first to cavitate within an annual ring, if the relationship between P_{50} and density is valid for wood at a given cambial age and if a similar relationship exists within a tree trunk.

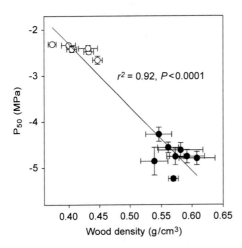

Figure 4. Clonal means of the negative of the air pressure applied by a means of a pressure collar causing 50 % loss of hydraulic conductivity (P_{50}) plotted against wood density of juvenile sapwood specimen (filled symbols, annual rings 2–3) of the trunk from 5-year-old trees and mature sapwood specimen (open symbols, annual rings 17–19) from the lower tree trunk of 24-year-old trees. Error bars show one standard error. Description of the study sites and methodological approaches for assessment of P_{50} can be found in Rosner *et al.* (2007) for juvenile wood and in Rosner *et al.* (2008) for mature wood. Wood density was calculated from the volume and the weight of the specimen in the oven-dried state.

Which are the hydraulically most vulnerable tracheids within an annual ring?

The efficiency of water transport increases with increasing conduit diameter (Pothier *et al.* 1989; Lo Gullo & Salleo 1991; Tyree *et al.* 1994) and as a direct function of variability in the resistance at the pit membrane (Pothier *et al.* 1989; Hacke *et al.* 2004, 2006). Within a conifer tree, hydraulic effectiveness and safety should be conflicting xylem functions: the most conductive conduits are supposed to be the most vulnerable ones (Sellin 1991; Cochard 1992; Hacke & Sperry 2001; Domec & Gartner 2001; Tyree & Zimmermann 2002). Norway spruce was the first species where two up-to-date non-destructive testing and analysis methods were applied from technical engineering in order to "look inside" dehydrating sapwood in order to test this hypothesis on the within-ring level. Acoustic emission (AE) feature analysis (Rosner *et al.* 2006) and neutron radiography (Rosner *et al.* 2012) gave quite new insights into processes associated with movement of free water inside Norway spruce sapwood.

The bulk of AE with the strongest frequencies in the range of 100–300 kHz are supposed to be induced by cavitation, *i.e.* the rapid tension release in the tracheid lumen as liquid water at negative pressure is suddenly replaced by water vapor (Milburn & Johnson 1966; Tyree *et al.* 1984; Kawamoto & Williams 2002). The plant physiological approach of AE testing has been focused on counting ultrasonic signals surpassing a defined detection threshold, on the assumption that the cumulated number of AE corresponds to a loss in hydraulic conductivity (*e.g.* Lo Gullo & Salleo 1991; Cochard 1992; Kikuta *et al.* 2003; Höltta *et al.* 2005). The energy of an acoustic signal, a waveform feature that depends on amplitude and signal duration, can give additional information on its source. Rosner *et al.* (2006) monitored radial dimensional changes together with AE of small dehydrating Norway spruce sapwood beams and performed post-hoc waveform analyses. The mean energies of acoustic signals plotted against time showed a typical pattern (Fig. 5a, d), where high energy AE is produced not right at the beginning but quite early during dehydration. According to the Hagen-Poiseuille equation, cavitation of big diameter tracheids causes much higher hydraulic losses than cavitation of small diameter tracheids. AE feature analysis (Rosner *et al.* 2006) allows quantifying cavitation events relative to their consequence upon conductivity loss, because AE energy or amplitude increases with increasing tracheid lumen area (Rosner *et al.* 2009; Mayr & Rosner 2011; Wolkerstorfer *et al.* 2012). It is hypothesized that bigger tracheids emit stronger AE signals due to cavitation because they are more prone to deformation upon negative pressure than smaller tracheids (Rosner *et al.* 2009). Dimensional changes in association with negative pressure in cell lumen or capillaries can be observed in sapwood of living stems (Neher 1993; Herzog *et al.* 1996; Zweifel *et al.* 2001; Cochard 2001; Offenthaler *et al.* 2001; Höltta *et al.* 2005; Conejero *et al.* 2007) and also during drying of isolated sapwood specimens (Irvine & Grace 1997; Rosner *et al.* 2009; Hansmann *et al.* 2011). Radial dimensional changes of Norway spruce sapwood occur immediately after the drying process starts (Fig. 5b, e). The partial recovery is due to a stress release when it comes to the breakage of the water columns inside the tracheids. When moisture content approaches towards fibre saturation, *i.e.* when the cell lumen contains no longer free water, but cell walls are fully saturated with liquid (Skaar *et al.* 1988), the observed recovery process is however overlaid by cell wall shrinkage (Fig. 5c, f).

By means of neutron radiography it was validated that dimensional changes of small sapwood beams at moderate moisture loss are not only caused by drying and rewetting of surface layers (Rosner *et al.* 2012). Neutrons are more sensitive than X-ray beams to some major elements, such as hydrogen, and they are particularly suitable for investigating both wood structure of dry wood as well as free or bound water in the wood (Lehmann *et al.* 2001; Mannes *et al.* 2009; Sonderegger *et al.* 2010). An oven-dry wood specimen still contains about 6% hydrogen, which accounts for 90% of the attenuation of the neutrons (Mannes *et al.* 2009). Neutron radiography is therefore suitable to investigate structure together with moisture distribution within a wood specimen

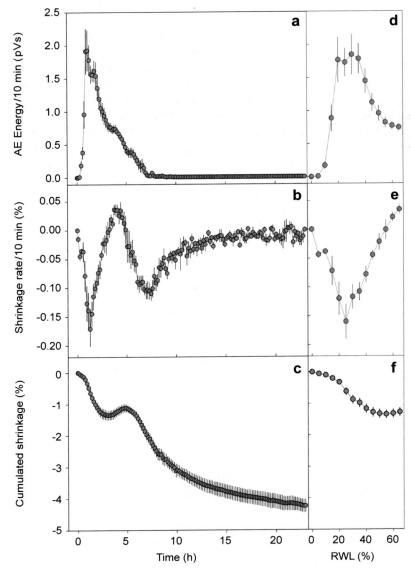

Figure 5. Courses of the mean acoustic emission energy/10 min (**a, d**), the rate of radial dimensional changes/10 min (**b, e**) and the cumulated radial dimensional changes (**c, f**) of dehydrating fresh mature Norway spruce sapwood beams (6 × 6 × 100 mm) plotted against time and against 5% relative water loss steps. Dehydration was performed at ambient conditions (25°C, 30% r.h.). Radial dimensional changes and acoustic emission were assessed with a load cell and a resonant 150 kHz acoustic transducer positioned on the tangential wood surface of a fully saturated specimen by means of a self-designed acrylic resin clamp assemblage which is described in detail in Rosner *et al.* (2009). Error bars (n = 12) show one standard error.

(Sonderegger *et al.* 2010). By means of this imaging technique, Rosner *et al.* (2012) could prove that free water in the middle part of a small Norway spruce sapwood beam is not continuously removed from the outer to the inner parts but earlier from low density earlywood than from the main latewood parts. In the transmission profiles (Fig. 6), moisture loss is indicated by an increase in transmission. Quite unexpected was, that

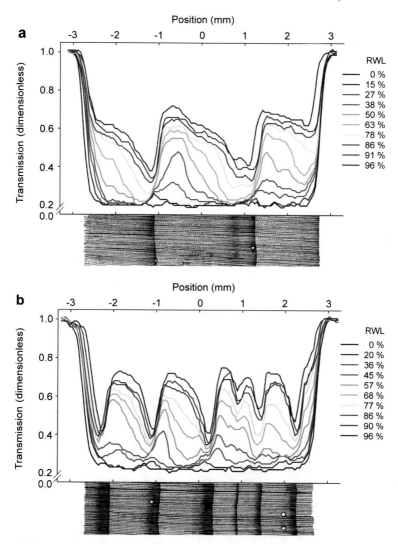

Figure 6. Transmission profiles of a Norway spruce sapwood beam with wide annual rings (**a**) and narrow annual rings (**b**) at different stages of relative moisture loss (RWL). Below each transmission profile the transverse section of the wood beam with dimensions at full saturation is shown. The x-axis represents the positions of the transmission profiles from both sides of the middle part of the beam (= 0 mm), since the specimen shows dimensional changes upon dehydration. The figure is modified after Rosner *et al.* (2012), where the methodological approach is described in detail.

not all latewood tracheids were less prone to cavitation, since transmission profiles gave strong evidence that some latewood tracheids lost free moisture earlier than earlywood tracheids. In the middle part of the specimens, where shrinkage is negligible (position around 0 mm), transmission after 20 % relative moisture loss was higher in latewood than in earlywood (Fig. 6b). Domec and Gartner (2002b) and Domec et al. (2006) reported similar findings for Douglas-fir sapwood. Latewood-bordered pits of Norway spruce do not aspirate upon dehydration (Liese & Bauch 1967) giving strong evidence that different mechanisms of air seeding exist in earlywood and latewood (Rosner 2012). The rigidity of the pit membrane is supposed to have an impact upon the mechanism how air seeding occurs (Gartner 1995; Domec & Gartner 2002b; Hacke et al. 2004; Domec et al. 2006, 2008). Pits act as valves, preventing the spread of bubbles through the conducting system (Tyree & Zimmermann 2002). Entry of air occurs either when the torus is not tightly sealed against the overarching pit border or when air bubbles are pulled through small pores in the torus or, in non-aspirated pits, through the margo of the pit membrane (Sperry & Tyree 1990; Tyree & Zimmermann 2002; Domec & Gartner 2002b; Choat & Pitterman 2009; Cochard et al. 2009; Delzon et al. 2010; Jansen et al. 2012). The strategy of Norway spruce earlywood to prevent air seeding is thus assumed to be pit aspiration (Fig. 2b), whereas latewood tracheids invest in rigid pit membranes with thicker margo strands and smaller pores. Air seeding in latewood will thus take place directly through the margo as hypothesized for Douglas fir (Domec & Gartner 2002b). Some latewood tracheids, probably those with the smallest lumen diameters, should however remain conductive even at low negative pressures, since they bear the smallest margo pores (Domec et al. 2006). To sum up, Norway spruce earlywood is more prone to cavitation than the main latewood parts. Some latewood tracheids are however highly vulnerable to cavitation, probably because their pits do not aspirate and pit membranes are not densely enough structured to avoid air entry. Whether cavitation of such tracheids has a high impact on conductivity loss should be a topic of further research.

Hydraulics and mechanics of the mature trunkwood: mean ring density is a predictive trait

Rosner et al. (2008) investigated the impact of growth and basic density on hydraulics and mechanical properties of six different Norway spruce clones. For the study, clones with different diameter and height growth were selected. The same clones were grown on two sites with different water availability in southern Sweden. Basic density, hydraulic, and mechanical parameters in mature spruce wood varied considerably between clones (Fig. 7), suggesting high breeding potential for these parameters (Rozenberg & Cahalan 1997; Hannrup et al. 2004). Stem wood of rapidly growing clones had significantly lower basic density, was more vulnerable to cavitation (Fig. 7a) and had higher values of sapwood area specific hydraulic conductivity at full saturation (Fig. 7b). Rapidly growing clones produced however trunkwood with lower bending strength and stiffness (modulus of elasticity, Fig. 7c) and compression strength and stiffness in the axial direction (Young's modulus). In accordance with Hacke et al. (2001), a clear tradeoff existed between hydraulic conductivity and vulnerability to cavitation, where spruce clones with high basic density had significantly lower hydraulic vulnerability but also lower hydraulic conductivity at full saturation and thus less rapid growth. As mentioned above, the relationship between hydraulic performance and density is not a direct one, but depends on the characteristics of the bordered pits.

Density is a quite good indirect predictive trait for hydraulic vulnerability (Fig. 7d), hydraulic conductivity at full saturation (Fig. 7e) and bending stiffness (Fig. 7f) in mature spruce wood. However, empirical models for predicting hydraulics or mechanics from basic density have to take account for site effects or forestry practices influencing

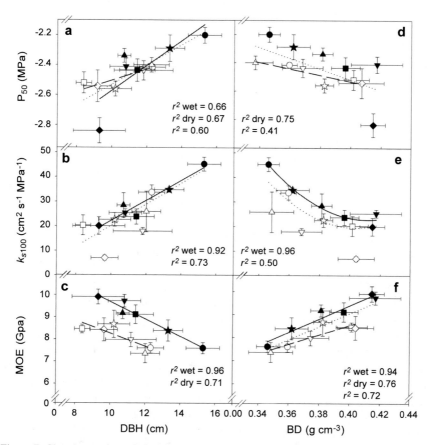

Figure 7. Clonal means of the diameter at breast height (DBH) and basic density (BD) as functions of clonal means of the negative of the air pressure applied by a means of a pressure collar causing 50% loss of hydraulic conductivity (P_{50}), the specific hydraulic conductivity at full saturation (k_s100), and the bending stiffness (MOE). Filled symbols denote trees from the wetter site (Tönnersjöheden); open symbols denote trees from the drier site (Vissefjärda); each symbol denotes a different clone. Error bars represent one standard error. Significant linear (a, b, c, d, f) and quadratic (e) relationships ($P < 0.05$) are indicated by solid regression lines for the wet site ($n = 6$) and by dashed regression lines for the dry site ($n = 6$). Dotted lines indicate significant linear (a, b, c, d, f) and quadratic (e) relationships across sites ($n = 12$). The figure is modified after Rosner *et al.* (2008), where detailed descriptions of clones, sites and methods can be found.

the physiological stage of the individual tree, *e.g.* of different crown lengths or stem taper (Deleuze *et al.* 1996; Kantola & Mäkelä 2004; Jaakkola *et al.* 2005). Although maximum hydraulic conductivity was negatively related to mechanical strength and stiffness on both sites and across sites, Rosner *et al.* (2008) found hints for smart structural solutions for achievement of both high hydraulic efficiency (the sapwood area-specific conductivity) and mechanical strength: clones growing on the wetter site showed significantly higher hydraulic conductivities and thus higher growth (Fig. 7e), but also higher mechanical strength and stiffness (Fig. 7f) than those on the drier site. Higher mechanical stiffness was not achieved by increasing mean ring density. Hydraulic efficiency and mechanical stability in the axial direction can be obviously obtained –to some extent– simultaneously. Such strategies are described in the chapter "Structural tradeoffs due to different hydraulic and mechanical demands within the trunk".

The genetic determination of wood density in Norway spruce is however generally high (Rozenberg & Cahalan 1997; Hannrup *et al.* 2004; Jaakkola *et al.* 2005). It is therefore suggested that also mechanical properties and vulnerability to cavitation might be under strong genetic control (Rosner *et al.* 2008). In species such as Norway spruce with growth rates inversely related to specific gravity (Herman *et al.* 1998; Hannrup *et al.* 2004; Steffenrem *et al.* 2009), rapid growth rate is the principal criterion for tree breeding and high density only second (Zobel & Jett 1995). Selecting for growth may thus lead not only to a reduction in mechanical strength and stiffness but also to a reduction in hydraulic safety if adequate precautions are not taken (Booker & Sell 1998; Domec & Gartner 2002a; Rozenberg *et al.* 2002; Cochard *et al.* 2007).

Hydraulic and mechanical demands along a tree trunk: Compromises are necessary

Natural selection has optimized wood structure within tree trunks to maximize survival of the species. The optimum structures for each biological wood function will most likely differ, leading to conflicting demands on wood structure for physiological fitness (Schniewind 1962; Baas 1983; Gartner 2001). Within a conifer tree trunk, structural, hydraulic and biomechanical characteristics show therefore a higher variability than can be found in mature trunkwood of different individuals from the same species summarized as the concept of juvenile wood, or more correct "core wood" (Gartner 1995; Lachenbruch *et al.* 2011). Juvenile Norway spruce wood is produced from cambial zones younger than 15–20 years (Kučera 1994; Saranpää 1994; Lindström *et al.* 1998) and is characterized by shorter cells with thinner cell walls and smaller lumen diameters (Fig. 3a,d), larger microfibril angles (the deviation of the microfibrils in the S_2 layer from the tracheid axis), different specific gravity and by within-ring density variations (Fig. 3b,c) compared to mature wood (Saranpää 1994; Lindström 1997; Saranpää *et al.* 2000; Sarén *et al.* 2001; Jungnikl *et al.* 2008). Kučera (1994) found that the formation of mature Norway spruce wood on the lower trunk commences when the annual height growth has culminated.

Mechanical demands within a conifer trunk with special reference to Norway spruce

Tree trunks experience short- and long-term mechanical stresses from a variety of causes such as gravity, wind, weight of snow, removal of a branch, partial failure of the anchorage system, or growth and development (Gartner 1995; Lachenbruch *et al.*

2011). The weight of a tree will subject the trunk largely to bending, but winds will normally cause twisting as well, because neither has a tree a completely symmetrical shape, nor is wind load uniform (Vogel 1995). Growth stress in the older wood develops along the tree axes in radial, tangential and axial direction due to sequential production of annual rings (Archer 1986; Gartner 1995). Where stems are out of their equilibrium positions due to static and dynamic forces of gravity and wind loads, compression wood is formed on the lower side of leaning stems or opposite to the windward side (Archer 1986; Timell 1986; Telewski 1995; Mattheck 1998; Spatz & Bruechert 2000). Spiral grain, *i.e.* the angle between stem axis and the inclination of the longitudinal tracheids in the direction of the wind-induced torque offers an additive advantage of the tree to cope with heavy wind loads (Skatter & Kučera 1997). Spiral-grained Norway spruce stems bend and twist more when exposed to strong winds, offering less wind resistance and being less likely to break. Moreover, through the bending and twisting, snow can slide down from branches rather than breaking them (Kubler 1991). Structure-function relationships within the main trunk presented hereafter are dealing with normal or opposite knot-free Norway spruce wood and mechanical testing was performed along or normal to the grain.

Rosner and Karlsson (2011) investigated bending stiffness of the outer sapwood within the trunk of 24-year-old Norway spruce trees. Bending stiffness was found to be lower at the tree top than at the lower tree trunk (Fig. 8e). Higher bending stiffness at the base than towards the top should provide several advantages for the stability of a Norway spruce tree, *e.g.* when it interacts with strong winds (Kubler 1991; Lundström *et al.* 2007). Bending resistance is highest where it is most needed (Mattheck 1998), thus higher up in the more tapered stems (Milne & Blackburn 1989). High flexibility of the upper part prevents the crown from catching wind flow and heavy snow loads are allowed to slip down from the crown rather than breaking it (Kubler 1991).

Hydraulic demands within Norway spruce trunks

Within a conifer trunk, hydraulic efficiency has been reported to increase from pith to bark, *i.e.* from juvenile to mature wood (Domec & Gartner 2001; Spicer & Gartner 2001; Domec & Gartner 2002a; Domec *et al.* 2009). Generally, wood higher up in the trunk (higher amount of juvenile wood) is modified to be more resistant to cavitation than the wood at the base because more negative pressures develop higher up in the crown (Cochard 1992; Domec & Gartner 2001; Domec & Gartner 2002a; Domec *et al.* 2009). These relationships proved valid as well for Norway spruce trunks; young trunkwood from the tree top had lower hydraulic vulnerability (Fig. 8a) as well as lower specific hydraulic conductivity than mature trunkwood of the lower trunk (Rosner *et al.* 2006; Rosner & Karlsson 2011; Wolkerstorfer *et al.* 2012). Higher hydraulic efficiency was achieved by higher hydraulic lumen diameters (Anfodillo *et al.* 2005, 2013; Domec *et al.* 2009) and most likely by modifications of the characteristics of bordered pits (Domec *et al.* 2006, 2008). Higher hydraulic vulnerability is aligned with lower wall/lumen rations in earlywood tracheids (Fig. 8a) (Domec *et al.* 2009), thus with a higher susceptibility to radial deformation (Fig. 8c) under a given (negative) pressure (Hacke *et al.* 2001; Hacke & Sperry 2001; Rosner & Karlsson 2011).

In accordance with Domec *et al.* (2009), who performed investigations on Douglas fir and Ponderosa pine, hydraulic vulnerability was not related to mean ring density within the main Norway spruce trunk: juvenile wood from the tree top showed similar density values as mature wood at breast height (Rosner & Karlsson 2011). Moreover, in pine species mean ring density was found to be even lower in juvenile than in mature trunkwood (Domec *et al.* 2009).

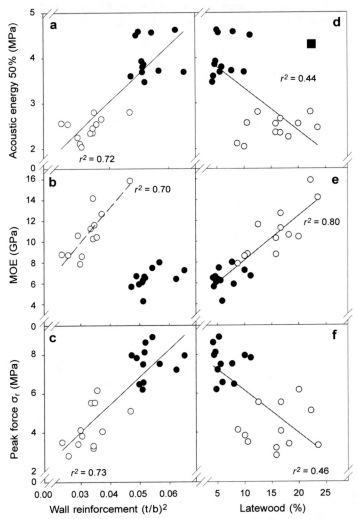

Figure 8. Hydraulic and mechanical traits related to anatomical properties. Vulnerability to cavitation assessed by the acoustic method (**a, d**), modulus of elasticity in bending (**b, e**) and compression strength (**c, f**) plotted against the conduit wall reinforcement and latewood percentage of juvenile (closed circles) and mature (open circles) Norway spruce sapwood specimens. Significant linear relationships across cambial age are indicated by solid regression lines, the significant relationship in mature wood by a broken regression line. Material and methods are described in detail in Rosner & Karlsson (2011).

Structural tradeoffs due to different hydraulic and mechanical demands within the trunk

An increase in density goes along with an increase in mechanical strength but also with a decrease in hydraulic efficiency, because it is achieved mainly by narrowing lumen diameter rather than increasing cell wall thickness (Hannrup *et al*. 2004; Pitterman *et al*. 2006; Sperry *et al*. 2006). Bending strength and axial compression strength, however, can be modified by varying the percentage and density of latewood (Hacke & Sperry 2001; Pitterman *et al*. 2006; Sperry *et al*. 2006; Jyske *et al*. 2008; Domec *et al*. 2009), fibre length (Mencuccini *et al*. 1997; Ezquerra & Gil 2001), arrangement of the cell wall layers (Jagels *et al*. 2003; Jagels & Visscher 2006), microfibril angle (Meylan & Probine 1969; Booker & Sell 1998; Lichtenegger *et al*. 1999; Evans & Ilic 2001; Jungnikl *et al*. 2008; Salmén & Burgert 2009) or cell wall chemistry (Gindl 2001, Gindl 2002; Kukkola *et al*. 2008). Moreover, hydraulic efficiency can be increased by increasing the pit pore size in earlywood (Mayr *et al*. 2002; Rosner *et al*. 2007). Within a tree trunk, hydraulic efficiency and bending strength as well as axial compression strength are therefore not necessarily conflicting wood functions (Mencuccini *et al*. 1997; Jagels & Visscher 2006). Accordingly, mature Norway spruce trunkwood showed higher hydraulic efficiency (hydraulic conductivity at full saturation) and bending stiffness than juvenile wood from the tree top, although both wood types had similar mean ring densities (Rosner & Karlsson 2011). Mean ring density in Norway spruce decreases till annual ring eight and increases again thereafter (Kučera 1994), which is most likely achieved by variations in latewood percentage or latewood density (Mäkinen *et al*. 2002; Jyske *et al*. 2008). Accordingly, high bending stiffness across cambial age within Norway spruce trunks was strongly related to higher latewood percentage (Fig. 8e). An increase with tracheid length goes along with an increase in bending stiffness as well as in hydraulic efficiency (Mencuccini *et al*. 1997; Ezquerra & Gil 2001) and is known to increase with cambial age in Norway spruce (Schultze-Dewitz 1959; Saranpää 1994; Lindström 1997; Sirviö & Kärenlampi 2001). High microfibril angles in the S_2 layer of the cell wall of juvenile Norway spruce wood (Lindström *et al*. 1998; Brändström 2001; Hannrup *et al*. 2004) guarantee higher flexibility of the upper crown (Neagu *et al*. 2006) but should have no negative implication upon vulnerability to cavitation. However, does Norway spruce trunkwood also manage to guarantee high mechanical strength perpendicular to the grain and high hydraulic efficiency?

Rosner and Karlsson (2011) were the first to relate compression strength perpendicular to the grain to anatomical and hydraulic parameters within a conifer trunk. Hydraulically less safe mature Norway spruce sapwood from the older trunk had higher specific hydraulic conductivity and bending stiffness (Fig. 8e), but lower radial compression strength (Fig. 8c) and conduit wall reinforcement (Fig. 8b) than hydraulically safer juvenile wood from the young crown. A clear tradeoff existed between hydraulic efficiency and compression strength perpendicular to the grain, since both traits are strongly related to the characteristics of the "weakest" wood parts, the low density earlywood (Müller *et al*. 2003). Structural compromises such as increasing conduit wall reinforcement in mature earlywood would probably be too costly for the tree. Radial compression strength (defined as the first peak in the stress/strength curve) is therefore negatively correlated to bending stiffness within a Norway spruce trunk (Rosner & Karlsson 2011). It is proposed that radial compression strength could be a highly predictive parameter for the resistance against vulnerability to cavitation, based on the close relationship to conduit wall reinforcement (Fig. 8c). These relationships should be tested in further studies either within one species or across species.

Hydraulic and mechanical demands of young Norway spruce stems

A young/small tree is in competition with others for light, water and nutrient supply but has a shallow root system and little water storage capacity (Lachenbruch *et al.* 2011; Scholz *et al.* 2011) which implies the need for higher hydraulic safety factors than are necessary in mature trunkwood of old-growth trees (Domec *et al.*2009). In species such as Norway spruce that manage to survive at the alpine timberline (Mayr *et al.* 2002, 2003) or in northern regions (Tollefsrud *et al.* 2008), the young trunks must as well be highly flexible in order to carry heavy wind and snow loads. Slight changes in stem orientation due to competing for light together with the important biomechanical task of keeping the stem in the upright position result in variable amounts of compression wood (Lindström *et al.* 1998; Zobel & Sprague 1998; Lachenbruch *et al.* 2011). Compression wood, usually present on the lower side of leaning stems or opposite to the windward side, has thicker tracheid cell walls and smaller tracheid lumen than opposite or "normal "wood. Tracheids of compression wood have a more rounded shape in the transverse plane and their cell walls contain more lignin and have higher S_2 microfibril angles than those of "normal" wood (Timell 1986; Gorisek & Torelli 1999; Bergander *et al.* 2002; Burgert *et al.* 2002; Gindl 2002). Norway spruce compression wood has a much higher density than normal wood but has also a higher vulnerability to cavitation (Mayr & Cochard 2003). Structure-function relationships are thus much more complex in the trunk of young/small spruce trees than in (normal) mature trunkwood because a density based tradeoff in hydraulic functions can be masked by the individual mechanical demands (Mayr *et al.* 2003).

Rosner *et al.* (2007) investigated structure-function relationships in 2–3-year-old stem segments from eight different Norway spruce clones (field age five years) differing in growth characteristics. Ring width, wood density, latewood percentage, lumen diameter, wall thickness, tracheid length and pit dimensions of earlywood cells, spiral grain and microfibril angles were tested together with hydraulic and mechanical traits. Vulnerability to cavitation (P_{50}, Fig. 4) and specific hydraulic conductivity of trunkwood from trees younger than five years were found to be much lower compared to mature trunkwood or wood from the tree top of older spruce trees (Rosner *et al.* 2006), which might be explained by limited access to ground water due to a less efficient root system. Other than in mature trunkwood, wood density was not related to the hydraulic vulnerability parameters assessed (Mayr *et al.* 2003), which comprised P_{12} and P_{50}, the pressure that is necessary to result in 12 % and 50 % loss of hydraulic conductivity, respectively, and PLC_{4MPa}, the percent loss of conductivity induced by 4 MPa pressure. Traits associated with higher hydraulic vulnerability in the young tree trunk were long tracheids (P_{12}) and thick earlywood cell walls (PLC_{4MPa}). The positive relationship between earlywood wall thickness and vulnerability to cavitation suggests that air seeding through the margo of the bordered pits may also occur in juvenile Norway spruce earlywood (Domec & Gartner 2002b; Jansen *et al.* 2012). Pit membranes of earlywood cells with thicker walls may be less flexible, so they cannot be that easily deflected to seal off the pit aperture completely (Sperry & Tyree 1990; Domec *et al.* 2006, 2008) as it is characteristic for compression wood (Mayr & Cochard 2003) and conifer latewood (Domec & Gartner 2002b).

One of the most interesting findings of Rosner *et al.* (2007) was that maximum hydraulic conductivity in young Norway spruce trunks is not only positively related to tracheid length and pit dimensions (Mayr *et al.* 2002) but also strongly to spiral grain. Spiral grain might offer an additional advantage for reducing flow resistance of the bordered pits of the first formed earlywood tracheids, which are characterized by rounded tips and a quite uniform distribution of pits along the entire length (Fig. 1a, b), as found in light bands of branch compression wood (Mayr *et al.* 2005). In earlywood formed later in the growing season, bordered pits achieving axial water flow can be found exclusively near the tapered tracheid ends (Fig. 1c). Hydraulic conductivity and vulnerability to cavitation, estimated as PLC_{4MPa}, showed only a weak tradeoff; both traits reached however higher values in trees with fast growth.

Variability in mechanical properties (bending strength and stiffness, axial compression strength and stiffness) of the young tree trunk depended mostly on wood density, but also on the amount of compression wood (Rosner *et al.* 2007). A density-based tradeoff between hydraulic characteristics and mechanical strength or stiffness (Hacke *et al.* 2001; Hacke & Sperry 2001; Domec & Gartner 2002a; Bouffier *et al.* 2003) is probably masked by structural compromises associated with mechanical demands of the young trunk (Mayr *et al.* 2002). High wood density in normal wood or compression wood can be compensated for flow reduction by a higher pit frequency in the first formed earlywood tracheids (Mayr *et al.* 2005). As mentioned above, mechanical support can be achieved not only by increasing mean ring wood density. Moreover, a proposed density-based tradeoff in hydraulic parameters is additionally masked by the fact that compression wood has a higher density than normal wood but is more vulnerable to cavitation (Mayr & Cochard 2003; Mayr *et al.* 2005). Mayr *et al.* (2003) measured in leader shoots of small Norway spruce trees from the alpine timber line a 1.4 times higher specific hydraulic conductivity as well as a 4.9 fold higher leaf specific conductivity than in side twigs, although vulnerability to cavitation was much lower in the former. They also found that lower vulnerability to cavitation is not related to wood density, expressed as the wall/lumen ratio (Hacke *et al.* 2001; Hacke & Sperry 2001), but correspond to smaller pit dimensions.

Genetic determination of hydraulic vulnerability was found to be quite weak in young stem segments, but its predictive structural traits were under strong genetic control (Rosner *et al.* 2007). Most of these traits, such as tracheid length and pit aperture percentage, were positively related to growth. The positive relationship between density and hydraulic safety (Hacke & Sperry 2001) will become more apparent in mature Norway spruce wood (Rosner *et al.* 2008); early selection for high growth (Zobel & Jett 1995) could thus result in individuals with increased sensitivity to cavitation in the mature trunkwood (Domec & Gartner 2002a).

CONCLUSIONS AND OUTLOOK

New applications in tree physiological research such as neutron radiography and acoustic emission feature analysis together with assessment of dimensional changes during dehydration helped to fill some gaps in our knowledge on hydraulic vulnerability within an annual ring of Norway spruce trunkwood. A future task in this regard is to perform anatomical and chemical analyses of cell walls and pit membranes in order to understand why some latewood compartments are more vulnerable to cavitation than low density earlywood. It was also necessary to investigate to what extent cavitation of highly vulnerable latewood tracheids contributes to conductivity loss since in mature Norway spruce wood, mean ring density is linked to hydraulic performance. Relating selected parameters from within-ring X-ray density profiles to vulnerability traits was the next step in our search for easily assessable parameters for estimating hydraulic performance under drought stress.

Up till now we know that sapwood of faster growing Norway spruce clones is more sensitive to cavitation and to mechanical stresses along the grain due to lower mean ring density; however, refinement of empirical models is still necessary. A precondition for using wood density parameters to screen for high hydraulic safety is taking account for structural compromises associated with mechanical demands within a tree trunk, *e.g.* guaranteeing high flexibility of the tree top. Thus, only anatomical data assessed in tree rings of similar cambial age sampled at defined (relative) distance either from the tree top or from the ground will give reliable results on hydraulic performance.

Structural compromises and complex wood structure, where a density-based tradeoff between hydraulic characteristics and bending stiffness is masked by the presence of reaction (compression) wood is probably the main reason why density is no useful trait for predicting hydraulic vulnerability in small Norway spruce trees. Thus, more research on reliable predictive traits for hydraulic safety of young juvenile wood is needed.

We should be aware that Norway spruce faces an uncertain future under the influence of continued global warming. Gaining more knowledge on structure-function relationships within tree trunks will enable the development of easily assessable and fast methods to estimate a tree's susceptibility to drought stress and will thus help selecting more suitable provenances or clones. An important task should be as well the investigation of refilling processes in Norway spruce sapwood, since a recent study points out that embolism repair is a key trait for the characterization of the strategy of a species to cope with drought stress (Johnson *et al.* 2012).

ACKNOWLEDGEMENTS

This review was financed by the Austrian Science Fund (FWF): V146-B16.

REFERENCES

Andreassen K, Solberg S, Tveito OE & Lystad SL. 2006. Regional differences in climatic responses of Norway spruce (*Picea abies* (L.) Karst.) growth in Norway. For. Ecol. Managem. 222: 211–221.

Anfodillo T, Carraio V, Carrer M, Fior C & Rossi S. 2005. Convergent tapering of xylem conduits in different woody species. New Phytol. 169: 279–290.

Anfodillo T, Petit G & Crivellaro A. 2013. Axial conduit widening in woody species: a still neglected anatomical pattern. IAWA J. 34: 352–364.

Archer RR. 1986. Growth stresses and strains in trees. Springer Verlag, Berlin, New York.

Baas P. 1983. Ecological patterns in xylem anatomy. In: Givnish TJ (ed.), On the economy of plant form and function: 327–352. Cambridge University Press, London, New York, New Rochelle, Melbourne, Sydney.

Barnard DM, Meinzer FC, Lachenbruch B, McCulloh KA, Johnson DM & Woodruff DR. 2011. Climate-related trends in sapwood biophysical properties in two conifers: avoidance of hydraulic dysfunction through coordinated adjustments in xylem efficiency, safety and capacitance. Plant Cell Environm. 34: 643–654.

Bergander A, Brändström J, Daniel G & Salmén L. 2002. Fibril angle variability in earlywood of Norway spruce using soft rot cavities and polarization confocal microscopy. J. Wood Sci. 48: 255–263.

Bergh J, Lindera S & Bergström J. 2005. Potential production of Norway spruce in Sweden. For. Ecol. Managem. 204: 1–10.

Booker RE & Sell J. 1998. The nanostructure of the cell wall of softwoods and its functions in a living tree. Holz Roh- Werkst. 56: 1–8.

Bouffier LA, Gartner BL & Domec J-C. 2003. Wood density and hydraulic properties of ponderosa pine from the Willamette valley vs. the Cascade mountains. Wood Fiber Sci. 35: 217–233.

Brändström J. 2001. Micro- and ultrastructural aspects of Norway spruce tracheids: a review. IAWA J. 22: 333–353.

Bréda N, Huc R, Granier A & Dreyer E. 2006. Temperate forest trees and stands under severe drought: a review of ecophysiological responses, adaptation processes and long-term consequences. Ann. For. Sci. 63: 625–644.

Burgert, I, Keckes J, Frühmann K, Fratzl P & Tschegg SE. 2002. A comparison of two techniques for wood fibre isolation - Evaluation by tensile tests on single fibres with different microfibril angle. Plant Biol. 4: 9–12.

Choat B, Jansen S, Brodribb TJ, Cochard H, Delzon S, Bhaskar R, Bucci SJ, TS Feild TS, SM Gleason SM, UG Hacke UG, AL Jacobsen AL, F Lens F, H Maherali H, Martínez-Vilalta J, Mayr S, Mencuccini M, Mitchell PJ, Nardini A, Pitterman J, Pratt RB, Sperry JS, Westoby M, Wright IJ & Zanne AE. 2012. Global convergence in the vulnerability of forests to drought. Nature 491: 752–755.

Choat B & Pitterman J. 2009. New insights into bordered pit structure and cavitation resistance in angiosperms and conifers. New Phytol. 182: 557–560.

Ciais P, Reichstein M, Viovy N, Granier A, Ogee J, Allard V, Aubinet M, Buchmann N, Bernhofer C, Carrara A, Chevallier F, De Noblet N, Friend AD, Friedlingstein P, Grunwald T, Heinesch B, Keronen P, Knohl A, Krinner G, Loustau D, Manca G, Matteucci G, Miglietta F, Ourcival JM, Papale D, Pilegaard K, Rambal S, Seufert G, Soussana JF, Sanz MJ, Schulze ED, Vesala T & Valentini R. 2005. Europe-wide reduction in primary productivity caused by the heat and drought in 2003. Nature 437: 529–533.

Cochard H. 1992. Vulnerability of several conifers to air embolism. Tree Physiol. 11: 73–83.

Cochard H. 2001. A new validation of the Scholander pressure chamber technique based on stem diameter variations. J. Exp. Bot. 52: 1361–1365.

Cochard H, Casella E & Mencuccini M. 2007. Xylem vulnerability to cavitation varies among poplar and willow clones and correlates with yield. Tree Physiol. 27: 1761–1767.

Cochard H, Hölttä T, Herbette S, Delzon S & Mencuccini M. 2009. New insights into the mechanisms of water-stress-induced cavitation in conifers. Plant Physiol. 151: 949–954.

Conejero W, Alarcón JJ, Garcia-Orellana Y & Nicolas E. 2007. Evaluation of sap and trunk diameter sensors for irrigation scheduling in early maturing peach trees. Tree Physiol. 27: 1753–1759.

Dalla-Salda G, Martinez-Meier A, Cochard H & Rozenberg P. 2009. Variation of wood density and hydraulic properties of Douglas-fir (*Pseudotsuga menziesii* (Mirb.) Franco) clones related to a heat and drought wave in France. For. Ecol. Managem. 257: 182–189.

Dalla-Salda G, Martinez-Meier A, Cochard H & Rozenberg P. 2011. Genetic variation of xylem hydraulic properties shows that wood density is involved in adaptation to drought in Douglas-fir (*Pseudotsuga menziesii* (Mirb.) Franco). Ann. For. Sci. 68: 747–757.

Deleuze C, Hervé J-C, Colin F & Ribeyrolles L. 1996. Modeling crown shape of *Picea abies*. Spacing effects. Can. J. For. Res. 26: 1957–1966.

Delzon S, Douthe C, Sala A & Cochard H. 2010. Mechanism of water-stress induced cavitation in conifers: bordered pit structure and function support the hypothesis of seal capillary-seeding. Plant Cell Environm. 33: 2101–2111.

Domec J-C. 2011. Let's not forget the critical role of surface tension in xylem water relations. Tree Physiol. 31: 359–360.

Domec J-C & Gartner BL. 2001 Cavitation and water storage in bole segments of mature and young Douglas-fir trees. Trees 15: 204–214.

Domec J-C & Gartner BL. 2002a. Age- and position-related changes in hydraulic versus mechanical dysfunction of xylem: inferring the design criteria for Douglas-fir wood structure. Tree Physiol. 22: 91–104.

Domec J-C & Gartner BL. 2002b. How do water transport and water storage differ in coniferous earlywood and latewood? J. Exp. Bot. 53: 2369–2379.

Domec J-C & Gartner BL. 2003. Relationship between growth rates and xylem hydraulic characteristics in young, mature and old-growth ponderosa pine trees. Plant Cell Environm. 26: 471–483.

Domec J-C, Lachenbruch BL & Meinzer FC. 2006. Bordered pit structure and function determine spatial patterns of air-seeding thresholds in xylem of Douglas-fir (*Pseudotsuga menziesii*; Pinaceae) trees. Amer. J. Bot. 93: 1600–1610.

Domec J-C, Lachenbruch BL, Meinzer FC, Woodruff DR, Warren JM & McCulloh KA. 2008. Maximum height in a conifer is associated with conflicting requirements for xylem design. Proc. Nat. Acad. Sci. USA 105: 12069–12074.

Domec J-C, Pruyn M & Gartner BL. 2005. Axial and radial profiles in conductivities, water storage, and native embolism in trunks of young and old-growth ponderosa pine trees. Plant Cell Environm. 28: 1103–1113.

Domec J-C, Warren JM, Meinzer FC & Lachenbruch BL. 2009. Safety for xylem failure by implosion and air-seeding within roots, trunks and branches of young and old conifer trees. IAWA J. 30: 101–120.

Dunham SM, Lachenbruch B & Ganio LM. 2007. Bayesian analysis of Douglas-fir hydraulic architecture at multiple scales. Trees 21: 65–78.

Evans R & Ilic J. 2001. Rapid prediction of wood stiffness from microfibril angle and density. For. Prod. J. 51: 53–57.

Ezquerra FJ & Gil LA. 2001. Wood anatomy and stress distribution in the stem of *Pinus pinaster* Ait. Invest. Agr.: Sist. Recur. For. 10: 165–177.

Gartner BL. 1995. Patterns of xylem variation within a tree and their hydraulic and mechanical consequences. In: Gartner BL (ed.), Plant stems: physiology and functional morphology: 125–149. Academic Press, San Diego.

Gartner BL. 2001. Multitasking and tradeoffs in stems, and the costly dominion of domatia. New Phytol. 151: 311–313.

Geßler A, Keitel C, Kreuzwieser J, Matyssek R, Seiler W & Rennenberg H. 2007. Potential risks for European beech (*Fagus sylvatica* L.) in a changing climate. Trees 21: 1–11.

Gindl W. 2001. The effect of lignin on the moisture-dependent behavior of spruce wood in axial compression. J. Mat. Sci. Letters 20: 2161–2162.

Gindl W. 2002. Comparing mechanical properties of normal and compression wood in Norway spruce: The role of lignin in compression parallel to the grain. Holzforschung 56: 395–401.

Gorisek Z & Torelli N. 1999. Microfibril angle in juvenile, adult and compression wood of spruce and silver fir. Phyton (Austria) 39: 129–132.

Greaves H. 1973. Comparative morphology of selected sapwood species using the scanning electron microscope. Holzforschung 27: 80–88.

Gregory SC & Petty JA. 1973. Valve action of bordered pits in conifers. J. Exp. Bot. 24: 763–767.

Hacke UG & Sperry JS. 2001. Functional and ecological xylem anatomy. Persp. Plant Ecol., Evolution Systematics 4: 97–115.

Hacke UG, JS Sperry & Pitterman J. 2004. Analysis of circular bordered pit function. II. Gymnosperm tracheids with torus-margo pit membranes. Amer. J. Bot. 91: 386–400.

Hacke UG, Sperry JS, Pockman WT, Davis SD & McCulloh K. 2001. Trends in wood density and structure are linked to prevention of xylem implosion by negative pressure. Oecologia 126: 457–461.

Hacke UG, Sperry JS, Wheeler JK & Castro L. 2006. Scaling of angiosperm xylem structure with safety and efficiency. Tree Physiol. 26: 689–701.

Hannrup B, Cahalan C, Chantre G, Grabner M, Karlsson B, Le Bayon I, Müller U, Pereira H, Rodrigues JC, Rosner S, Rozenberg P, Wilhelmsson L & Wimmer R. 2004. Genetic parameters of growth and wood quality traits in *Picea abies*. Scand. J. For. Res. 19: 14–29.

Hansmann C, Konnerth J & Rosner S. 2011. Digital image analysis of radial shrinkage of fresh spruce (*Picea abies* L.) wood. Wood Mat. Sci. Eng. 6: 2–6.

Havimo M., Rikala J, Sirviö J & Sipi M. 2008. Distributions of tracheid cross-sectional dimensions in different parts of Norway spruce stems. Silva Fennica 42: 89–99.

Herman M, Dutilleul P & Avella-Shaw T. 1998. Growth rate effects on temporal trajectories of ring width, wood density, and mean tracheid length in Norway spruce (*Picea abies* (L.) Karst.). Wood Fiber Sci. 30: 6–17.

Herzog KM, Häsler R & Thum R. 1996. Diurnal changes in the radius of a supalpine Norway spruce stem: their relation to the sap flow and their use to estimate transpiration. Trees 10: 94–101.

Hölttä T, Vesala T, Nikinmaa E, Perämäki M, Siivola E & Mencuccini M. 2005. Field measurements of ultrasonic acoustic emissions and stem diameter variations. New insight into the relationship between xylem tensions and embolism. Tree Physiol. 25: 237–243.

Hukin D, Cochard H, Dreyer E, Le Thiec D & Bogeat-Triboulot MB. 2005. Cavitation vulnerability in roots and shoots: does *Populus euphratica* Oliv., a poplar from arid areas of Central Asia, differ from other poplar species? J. Exp. Bot. 56: 2003–2010.

IPCC. 2012. Managing the risks of extreme events and disasters to advance climate change adaptation. A special report of working groups I and II of the Intergovernmental Panel on Climate Change. Field CB, Barros V, Stocker TF, Qin D, Dokken DJ, Ebi KL, Mastrandrea MD, Mach KJ, Plattner G-K, Allen SK, Tignor M & Midgley PM (eds.). Cambridge University Press, Cambridge, UK, and New York, NY, USA. 582 pp.

Irvine J & Grace J. 1997. Continuous measurements of water tensions in the xylem of trees based on the elastic properties of wood. Planta 202: 455–461.

Jaakkola T, Mäkinen H & Saranpää P. 2005. Effects of thinning and fertilisation on tracheid dimensions and lignin content of Norway spruce. Holzforschung 61: 301–310.

Jagels R & Visscher GE. 2006. A synchronous increase in hydraulic conductive capacity and mechanical support in conifers with relatively uniform xylem structure. Amer. J. Bot. 93: 179–187.

Jagels R, Visscher GE, Lucas J & Goodell B. 2003. Paleo-adaptive properties of the xylem of *Metasequoia*: Mechanical/hydraulic compromises. Ann. Bot. 92: 79–88.

Jansen S, Lamy J-B, Burlett R, Cochard H, Gasson P & Delzon S. 2012. Plasmodesmatal pores in the torus of bordered pit membranes affect cavitation resistance of conifer xylem. Plant Cell Environm. 35: 1109–1120.

Johnson DM, McCulloh KA, Woodruff DR & Meinzer FC. 2012. Hydraulic safety margins and embolism reversal in stems and leaves: Why are conifers and angiosperms so different? Plant Sci. 135: 4–53.

Jungnikl K, Koch G & Burgert I. 2008. A comprehensive analysis of the relation of cellulose microfibril orientation and lignin content in the S_2 layer of different tissue types of spruce wood (*Picea abies* (L.) Karst.). Holzforschung 62: 475–480.

Jyske T, Mäkinen H & Saranpää P. 2008. Wood density within Norway spruce stems. Silva Fenn. 42: 349–355.

Kantola A & Mäkelä A. 2004. Crown development in Norway spruce (*Picea abies* (L.) Karst.). Trees 18: 408–421.

Kapeller S, Lexer MJ, Geburek T, Hiebl J & Schueler S. 2012. Intraspecific variation in climate response of Norway spruce in the eastern Alpine range: Selecting appropriate provenances for future climate. Forest Ecol. Managem. 271: 46–57.

Kawamoto S & Williams RS. 2002. Acoustic emission and acousto-ultrasonic techniques for wood and wood-based composites – A Review. Madison, WI: Gen. Techn. Rep. FPL-GTR-134. U.S. Department of Agriculture, Forest Service, Forest Products Laboratory.

Kikuta SB, Hietz P & Richter H. 2003. Vulnerability curves from conifer sapwood sections exposed over solutions with known water potentials. J. Exp. Bot. 54: 2149–2155.

Kubler H. 1991. Function of spiral grain in trees. Trees 5: 125–135.

Kučera B. 1994. A hypothesis relating current annual height increment to juvenile wood formation in Norway spruce. Wood Fiber Sci. 26: 152–167.

Kukkola E, Saranpää P & Fagerstedt K. 2008. Juvenile and compression wood cell wall layers differ in lignin structure in Norway spruce and Scots pine. IAWA J. 29: 47–54.

Kullman L & Öberg L. 2009. Post-Little Ice Age tree line rise and climate warming in the Swedish Scandes: a landscape ecological perspective. J. Ecol. 97: 415–429.

Lachenbruch B, Moore J & Evans R. 2011. Radial variation in wood structure and function in woody plants, and hypotheses for its occurrence. In: Meinzer FC, Lachenbruch B & Dawson TE (eds.), Size and age-related changes in tree structure and function: 121–164. Springer, Dordrecht.

Lehmann E, Vontobel P, Scherrer P & Niemz P. 2001. Application of neutron radiography as method in the analysis of wood. Holz Roh- Werkst. 59: 463–471.

Lichtenegger H, Reiterer A, Stanzl-Tschegg SE & Fratzl P. 1999. Variation of cellulose microfibril angles in softwoods and hardwoods: A possible strategy of mechanical optimization. J. Struct. Biol. 128: 257–269.

Liese W & Bauch J. 1967. On the closure of bordered pits in conifers. Wood Sci. Techn. 1: 1–13.

Lindström H. 1997. Fiber length, tracheid diameter, and latewood percentage in Norway spruce: Development from pith outwards. Wood and Fiber Science 29: 21–34.

Lindström H, Evans JW & Verrill SP. 1998. Influence of cambial age and growth conditions on microfibril angle in young Norway spruce (*Picea abies* (L.) Karst.). Holzforschung 52: 573–581.

Lo Gullo MA & Salleo S. 1991. Three different methods for measuring xylem cavitation and embolism: a comparison. Ann. Bot. 67: 417–424.

Lundgren C. 2004. Cell wall thickness and tangential and radial cell diameter of fertilized and irrigated Norway spruce. Silva Fenn. 38: 95–106.

Lundström T, Heiz U, Stoffel M & Stöckli V. 2007. Fresh-wood bending: linking the mechanical and growth properties of a Norway spruce stem. Tree Physiol. 27: 1229–1241.

Maherali H, Pockman WT & Jackson RB. 2004. Adaptive variation in the vulnerability of woody plants to xylem cavitation. Ecology 85: 2184–2199.

Mäkinen H, Saranpää P & Linder S. 2002. Wood-density variation of Norway spruce in relation to nutrient optimization and fibre dimensions. Can. J. For. Res. 31: 185–194.

Mannes D, Sonderegger W, Hering S, Lehmann E & Niemz P. 2009. Non-destructive determination and quantification of diffusion processes in wood by means of neutron imaging. Holzforschung 63: 589–596.

Mattheck C. 1998. Design in nature: learning from trees. Springer Verlag, Berlin, Heidelberg, New York.

Mayr S, Bardage S & Brändström J. 2005. Hydraulic and anatomical properties of light bands in Norway spruce compression wood. Tree Physiol. 26: 17–23.

Mayr S & Cochard H. 2003. A new method for vulnerability analysis of small xylem areas reveals that compression wood of Norway spruce has lower hydraulic safety than opposite wood. Plant Cell Environm. 26: 1365–1371.

Mayr S & Rosner S. 2011. Cavitation in dehydrating xylem of *Picea abies*: energy properties of ultrasonic emissions reflect tracheid dimensions. Tree Physiol. 31: 59–67.

Mayr S, Rothart B & Dämon B. 2003. Hydraulic efficiency and safety of leader shoots and twigs in Norway spruce growing at the Alpine timberline. J. Exp. Bot. 54: 2563–2568.

Mayr S, Wolfschweger M & Bauer H. 2002. Winter-drought induced embolism in Norway spruce (*Picea abies*) at the Alpine timberline. Physiol. Plant. 115: 74–80.

Mayr S & Zublasing V. 2010. Ultrasonic emissions from conifer xylem exposed to repeated freezing. J. Plant Physiol. 167: 34–40.

McDowell N, Pockman WT, Allen CD, Breshears DD, Cobb N, Kolb T, Plaut J, Sperry J, West A, Williams DG & Ypez EA. 2008. Mechanisms of plant survival and mortality during drought: why do some plants survive while others succumb to drought? New Phytol. 178: 719–739.

Meinzer FC, Johnson DM, Lachenbruch B, McCulloh KA & Woodruff DR. 2009. Xylem hydraulic safety margins in woody plants: coordination of stomatal control of xylem tension with hydraulic capacitance. Functional Ecol. 23: 922–930.

Meinzer FC, McCulloh KA, Lachenbruch B, Woodruff DR & Johnson DM. 2010. The blind men and the elephant: The impact of context and scale in evaluating conflicts between plant hydraulic safety and efficiency. Oecologia 164: 287–296.

Mencuccini M, Grace J & Fioravanti M. 1997. Biomechanical and hydraulic determinants of tree structure in Scots pine: anatomical characterstics. Tree Physiol. 17: 105–113.

Meylan BA & Probine MC. 1969. Microfibril angle as a parameter in timber quality assessment. Forest Prod. J. 19: 31–34.

Milburn JA & Johnson RPC. 1966. The conduction of sap. I. Detection of vibrations produced by sap cavitation in *Ricinus* xylem. Planta 69: 43–52.

Milne R & Blackburn P. 1989. The elasticity and vertical distribution of stress within stems of *Picea sitchensis*. Tree Physiol. 5: 195–205.

Monclus R, Dryer E, Villar M, Delmotte FM, Delay D, Petit J-M, Barbaroux C, Le Thiec D, Bréchet C & Brignolas F. 2005. Impact of drought on productivity and water use efficiency in 29 genotypes of *Populus deltoides* x *Populus nigra*. New Phytol. 169: 765–777.

Müller U, Gindl W & Teischinger A. 2003. Effects of cell wall anatomy on the plastic and elastic behaviour of different wood species loaded perpendicular to grain. IAWA J. 24: 117–128.

Neagu RC, Gamstedt EK, Bardage SL & Lindström M. 2006. Ultrastructural features affecting mechanical properties of wood fibres. Wood Mat. Sci. Engin. 1: 146–170.

Neher V. 1993. Effects of pressures inside Monterey pine trees. Trees 8: 9–17.

Offenthaler I, Hietz P & Richter H. 2001. Wood diameter indicates diurnal and long-term patterns of xylem water potential in Norway spruce. Trees 15: 215–221.

Peuke AD, Schraml C, Hartung W & Rennenberg H. 2002. Identification of drought-sensitive beech ecotypes by physiological parameters. New Phytol. 154: 373–387.

Pitterman J, Sperry JS, Wheeler JK, Hacke UG & Sikkema EH. 2006. Mechanical reinforcement of tracheids compromises the hydraulic efficiency of conifer xylem. Plant Cell Environm. 29: 1618–1628.

Plomion C, Leprovost G & Stokes A. 2001. Wood formation in trees. Plant Physiol. 127: 1513–1523.

Pockman WT & Sperry JS. 2000. Vulnerability to xylem cavitation and the distribution of sonorian desert vegetation. Amer. J. Bot. 87: 1287–1299.

Pothier D, Margolis HA, Poliquin J & Waring RH. 1989. Relation between the permeability and the anatomy of jack pine sapwood with stand development. Can. J. For. Res. 19: 1564–1570.

Reich PB & Oleksyn J. 2008. Climate warming will reduce growth and survival of Scots pine except in the far north. Ecol. Letters 11: 588–597.

Richter H. 2001. The cohesion theory debate continues: the pitfalls of cryobiology. Trends in Plant Sci. 6: 456–457.

Rosner S. 2012. Waveform features of acoustic emission provide information about reversible and irreversible processes during spruce sapwood drying. BioResources 7: 1253–1263.

Rosner S & Karlsson B. 2011. Hydraulic efficiency compromises compression strength perpendicular to the grain in Norway spruce trunkwood. Trees 25: 289–299.

Rosner S, Karlsson B, Konnerth J & Hansmann C. 2009. Shrinkage processes in standard-size Norway spruce wood specimens with different vulnerability to cavitation. Tree Physiol. 29: 1419–1431.

Rosner S, Klein A, Müller U & Karlsson B. 2007. Hydraulic and mechanical properties of young Norway spruce clones related to growth and wood structure. Tree Physiol. 27: 1165–1178.

Rosner S, Klein A, Müller U & Karlsson B. 2008. Tradeoffs between hydraulic and mechanical stress response of mature Norway spruce trunkwood. Tree Physiol. 28: 1179–1188.

Rosner S, Klein A, Wimmer R & Karlsson B. 2006. Extraction of features from ultrasound acoustic emissions: a tool to assess the hydraulic vulnerability of Norway spruce trunkwood? New Phytol. 171: 105–116.

Rosner S, Konnerth J, Plank B, Salaberger D & Hansmann C. 2010. Radial shrinkage and ultrasound acoustic emissions of fresh *versus* pre-dried Norway spruce wood. Trees 24: 931–940.

Rosner S, Riegler M, Vontobel V, Mannes D, Lehmann E, Karlsson B & Hansmann C. 2012. Within-ring movement of free water in dehydrating Norway spruce sapwood visualized by neutron radiography. Holzforschung 66: 751–756.

Rozenberg P & Cahalan C. 1997. Spruce and wood quality: genetic aspects – A review. Silvae Genetica 46: 270–279.

Rozenberg P, Van Loo J, Hannrup B & Grabner M. 2002. Clonal variation of wood density record of cambium reaction to water deficit in *Picea abies* (L.) Karst. Ann. For. Sci. 59: 533–540.

Salinger S. 2005. Increasing climate variability and change: reducing the vulnerability. Clim. Change 70: 1–3.

Salmén L & Burgert I. 2009. Cell wall features with regard to mechanical performance. Holzforschung 63: 121–129.

Saranpää P. 1994. Basis density, longitudinal shrinkage and tracheid lenght of juvenile wood of *Picea abies* (L.) Karst. Scand. J. For. Res. 9: 68–74.

Saranpää P, Pesonen E, Saren M, Andersson S, Siiria S, Serimaa R & Paakkari T. 2000. Variation of the properties of tracheids in Norway spruce (*Picea abies* [L.] Karst). In: Savidge RA, Barnett JR & Napier R (eds.), Cell and molecular biology of wood formation: 337–345. BIOS Scientific Publishers, Oxford.

Sarén M-P, Serimaa R, Andersson S, Paakkari T, Saranpää P & Pesonen E. 2001. Structural variation of tracheids in Norway spruce (*Picea abies* (L.) Karst.). J. Struct. Biol. 163: 101–109.

Schär C, Vidale PL, Lüthi D, Frei C, Häberli C, Mark A, Lindinger MA & Appenzeller C. 2004. The role of increasing temperature variability in European summer heatwaves. Nature 427: 332–336.

Schlyter P, Stjernquist I, Bärring L, Jönsson AM & C Nilsson. 2006. Assessment of the impacts of climate change and weather extremes on boreal forests in northern Europe, focusing on Norway spruce. Clim. Res. 31: 75–84.

Schniewind AP. 1962. Horizontal specific gravity variation in tree stems in relation to their support function. Forest Sci. 8: 111–118.

Scholz FG, Phillips NG, Bucci SJ, Meinzer FC & Goldstein G. 2011. Hydraulic capacitance: biophysics and functional significance of internal water sources in relation to tree size. In: Meinzer FC, Lachenbruch B & Dawson TE (eds.), Size- and age-related changes in tree structure and function: 341–361. Springer, Dordrecht.

Schultze-Dewitz G. 1959. Variation und Häufigkeit der Faserlänge der Fichte. Holz Roh- Werkstoff 17: 319–326.

Sellin A. 1991. Hydraulic conductivity of xylem depending on water saturation level in Norway spruce (*Picea abies* (L.) Karst.). J. Plant Physiol. 138: 466–469.

Semenov VA. 2012. Arctic warming favours extremes. Nature Global Change 2: 315–316.

Sirviö J & Kärenlampi P. 1998. Pits as natural irregularities in softwood fibers. Wood Fiber Sci. 30: 27–39.

Sirviö J & Kärenlampi P. 2001. The effects of maturity and growth rate on the properties of spruce wood tracheids. Wood Sci. Techn. 35: 541–554.

Skaar C. 1988. Wood-water relations. Springer-Verlag, Berlin, Germany.

Skatter S & Kučera B. 1997. Spiral grain - An adaptation of trees to withstand stem breakage caused by wind-induced torsion. Holz Roh- Werkst. 55: 207–213.

Solberg S. 2004. Summer drought: a driver for crown condition and mortality of Norway spruce in Norway. Forest Pathology 34: 93–104.

Sonderegger W, Hering S, Mannes D, Vontobel P, Lehmann E & Niemz P. 2010. Quantitative determination of bound water diffusion in multilayer boards by means of neutron imaging. Eur. J. Wood Prod. 68: 341–350.

Spatz H-C & Bruechert F. 2000. Basic biomechanics of self-supporting plants: wind loads and gravitational loads on a Norway spruce tree. Forest Ecol. Managem. 135: 33–44.

Sperry JS 1995. Limitations on stem water transport and their consequences. In: Gartner BL (ed.), Plant stems: physiology and functional morphology: 105–124. Academic Press, San Diego.

Sperry JS, Hacke UG & Pitterman J. 2006. Size and function in conifer tracheids and angiosperm vessels. Amer. J. Bot. 93: 1490–1500.

Sperry JS & Ikeda T. 1997. Xylem cavitation in roots and stems of Douglas-fir and white fir. Tree Physiol. 17: 275–280.

Sperry JS, Meinzer FC & McCulloh KA. 2008. Safety and efficiency conflicts in hydraulic architecture: scaling from tissues to trees. Plant Cell Environm. 31: 632–645.

Sperry JS & Tyree MT. 1990. Water stress induced xylem cavitation in three species of conifers. Plant Cell Environm. 13: 427–436.

Spicer R & Gartner BL. 2001. The effects of cambial age and position within the stem on specific conductivity in Douglas-fir (*Pseudotsuga menziesii*) sapwood. Trees 15: 222–229.

Steffenrem A, Kvaalen H, Høibø OA, Edvardsen ØM & Skrøppa T. 2009. Genetic variation of wood quality traits and relationships with growth in *Picea abies*. Scand. J. Forest Res. 24: 15–27.

Telewski FW. 1995. Wind induced physiological and developmental responses in trees. In: Coutts MP & Grace J (eds.), Wind and trees: 237–263. Cambridge University Press, Cambridge.

Timell TE. 1986. Compression wood in conifers. Springer, Berlin, Heidelberg, New York.

Tollefsrud MM, Kissling R, Gugerli F, Johnsen Ø, Skrøppa T, Cheddadi R, Van der Knaap WO, Latałowa M, Terhürne-Berson R, Litt T, Geburek T, Brochmann C & Sperisen C. 2008. Genetic consequences of glacial survival and postglacial colonization in Norway spruce: combined analysis of mitochondrial DNA and fossil pollen. Molecular Ecol. 17: 4134–4150.

Tyree MT. 2003. The ascent of water. Nature 423: 923.

Tyree MT, Davis SD & Cochard H. 1994. Biophysical perspectives of xylem evolution: is there a tradeoff of hydraulic efficiency for vulnerability to dysfunction? IAWA J. 15: 335–360.

Tyree MT, Dixon MA & Thompson RG. 1984. Ultrasonic acoustic emissions from the sapwood of *Thuja occidentalis* measured inside a pressure bomb. Plant Physiol. 74: 1046–1049.

Tyree MT & Ewers FW. 1991. The hydraulic architecture of trees and other woody plants. New Phytol. 119: 345–360.

Tyree MT & Sperry JS. 1989. Vulnerability of xylem to cavitation and embolism. Ann. Rev. Plant Physiol. 40: 19–38.

Tyree MT & Zimmermann MH. 2002. Xylem structure and the ascent of sap. Ed. 2. Springer, Berlin, Germany.

Vogel S. 1995. Twist-to-bend ratios of woody structures. J. Exp. Bot. 46: 981–985.

Williams AP, Allen CD, Macalady AD, Griffin D, Woodhouse CA, Meko DM, Swetnam TW, Rauscher SA, Seager R, Grissino-Mayer HD, Dean JS, Cook ER, Gangodagamage C, Cai M & McDowell NG. 2012. Temperature as a potent driver of regional forest drought stress and tree mortality. Nature Climate Change 3: 292–297.

Wolkerstorfer SV, Rosner S & Hietz P. 2012. An improved method and data analysis for ultrasound acoustic emissions and xylem vulnerability in conifer wood. Physiol. Plant. 146: 185–191.

Zimmermann MH. 1983 Xylem structure and the ascent of sap. Springer, Berlin.

Zobel BJ & Jett JB. 1995. Genetics of wood production. Timell TE (ed.). Springer, Berlin, Heidelberg, New York.

Zobel BJ & Sprague JR. 1998. Juvenile wood in forest trees. Timell TE (ed.). Springer, Berlin, Heidelberg, New York.

Zobel BJ & Van Buijtenen JP. 1989. Wood variation: its causes and control. Timell TE (ed.). Springer, Berlin, Heidelberg, New York.

Zweifel R, Item H & Häsler R. 2001. Link between diurnal stem radius changes and tree water relations. Tree Physiol. 21: 869–877.

Accepted: 2 August 2013

IAWA Journal 34 (4), 2013: 391–407

BRILL

REVIEW OF CELLULAR AND SUBCELLULAR CHANGES IN THE CAMBIUM

Peter Prislan[1,2,*], Katarina Čufar[2], Gerald Koch[3], Uwe Schmitt[3] and Jožica Gričar[1]

[1]Slovenian Forestry Institute, Večna pot 2, SI-1000 Ljubljana, Slovenia
[2]Biotechnical Faculty, Department of Wood Science and Technology, University of Ljubljana, Rožna dolina, Cesta VIII/34, SI-1000 Ljubljana, Slovenia
[3]Thünen Institute of Wood Research, Leuschnerstraße 91, D-21031 Hamburg, Germany
*Corresponding author; e-mail: peter.prislan@gozdis.si

ABSTRACT

The commonest approach to studying cambial productivity is conventional light microscopy, which is widely used in wood formation studies. The number of such studies has increased rapidly in the past decade, usually in order to elucidate the relationship between growth and environmental factors. However, some aspects of cambial seasonality are often overlooked or neglected. Observations with transmission electron microscopy provide a more detailed insight into changes occurring on the ultra-structural level in cambial cells. Criteria for defining cambial activity are not yet fully clarified, especially when observing it at different resolutions, *i.e.*, on cellular, subcellular and ultrastructural levels. The goal of this review is to contribute to clarification of the terms mainly used, such as cambial dormancy, reactivation, activity, productivity and transition between different states, resting period and quiescence, which describe structural modifications of cambial cells during the various phases of their seasonal cycle. Based on our own cambium observations on adult beech trees growing at two different elevations, which were made with light and transmission electron microscopy, we discuss the influence of weather conditions on cambial activity and the advantage of the complementary use of different techniques and resolutions.

Keywords: Cambial activity, cambial cells, *Fagus sylvatica*, light microscopy, transmission electron microscopy, ultrastructural changes.

INTRODUCTION

Growth in plants takes place in specialised tissues, so-called meristems, which act as central control points for growth and development, receiving, integrating, responding to and broadcasting growth-regulating signals (Risopatron *et al.* 2010). Two meristems are responsible for the growth of trees: apical (growth in length) and lateral (growth in girth) (Lachaud *et al.* 1999). Apical meristems produce primary tissues, whereas lateral meristems (*i.e.*, vascular and cork cambium) contribute to the production of secondary tissues (Mauseth 2009). The vascular cambium develops from the procambium, which in turn is derived from parenchyma cells that have regained the capability to divide

© International Association of Wood Anatomists, 2013
Published by Koninklijke Brill NV, Leiden

DOI 10.1163/22941932-00000032

(Evert 2006). The cambium is a bifacial meristem because it produces xylem cells in the centripetal and phloem cells in the centrifugal direction (Larson 1994).

Cambial activity ensures the perennial life of trees through regular renewal of functional xylem and phloem (Plomion *et al.* 2001). Moreover, cambial growth might be considered to be the tree's way of ensuring that its stem and branches have sufficient structural support and hydraulic conductivity to grow against gravity, while also providing for the needs of the root system (Savidge 2000b). The annual course of cambial activity is generally related to the alternation of cold and warm or dry and rainy seasons (Lachaud *et al.* 1999). In some climatic regions, *i.e.* tropics, cambial activity may continue throughout the year, whereas in temperate regions the activity of the cambium is usually periodical, subjected to the tree's internal regulation (genetic and hormonal) (Ursache *et al.* 2013) and environmental factors, such as temperature, precipitation, photoperiod and other biotic and abiotic influences (*e.g.* Wodzicki 2001; Evert 2006; Begum *et al.* 2013).

Commonly, xylem production represents the major proportion of the tree's radial growth. In addition to the economic importance of wood, and partly also bark, wood increments provide an integral archive of factors affecting its formation before and/or during the time of cambial growth. Dendroclimatological and dendroecological studies, in combination with intra-annual observations of radial growth of trees, are thus useful in order better to understand climate-growth relationships (Čufar *et al.* 2008; Callado *et al.* 2013; Costa *et al.* 2013).

Although numerous investigations have been dedicated to the annual rhythm of cambial activity in various tree species, criteria for determining its activity are still not satisfactorily defined, especially in the case of observations (and their comparisons) at different levels, *e.g.*, cellular, sub-cellular and ultra-structural levels (Frankenstein *et al.* 2005; Prislan *et al.* 2011; Rathgeber *et al.* 2011). When observing ultrastructure of cambial cells, for example, the onset of cambial activity can be easily defined when first mitotic figures and phragmoplasts are observed (Larson 1994; Farrar & Evert 1997). However, changes in the ultrastructure of cells, which could also be considered as activity, occur before formation of phragmoplast, as a transition from the dormant to the active state (Farrar & Evert 1997). In wood formation studies, phragmoplasts or mitotic figures cannot be observed due to different methodologies of sample preparation and observation. Therefore reactivation of the cambium is defined by an increase in number of cambial cells and the occurrence of newly formed xylem and phloem cells in early developmental stages (*e.g.*, Gričar *et al.* 2006; Deslauriers *et al.* 2008). Rathgeber *et al.* (2011), who studied cambial phenology and xylem formation in *Abies alba*, pointed out that the number of cells in the cambium cannot be a precise indicator for its activity, because the increase in cell number may be insignificant. Consequently, the appearance of first enlarging xylem cells was considered as a more appropriate indicator for cambial reactivation (Rathgeber *et al.* 2011). In contrast, Frankenstein *et al.* (2005) observed that initial earlywood vessels in ring porous *Fraxinus excelsior* were formed in the previous growing season and overwintered, and then started to differentiate prior to the onset of cambial cell divisions in spring.

Aims of this review are to: (i) highlight the latest findings on structural modifications in cambium that are related to its seasonal activity and cell production; (ii) exemplify them by our most recent observations on seasonal cellular and ultrastructural changes in cambium (Prislan *et al.* 2011) and phenological variation in cambial productivity (Prislan *et al.* 2013) carried out in *Fagus sylvatica* growing under different weather conditions and (iii) emphasise the differences in observations at the cellular and ultra-structural level in order to suggest a suitable terminology for determining the onset of cambial activity at different observational levels.

STRUCTURE AND FUNCTION OF CAMBIUM

The cambium is composed of highly vacuolated meristematic cells organised in radial files, which give rise to the secondary xylem and phloem. Theoretically, each radial file contains one initial cell, which remains in the meristem, as well as phloem and xylem mother cells, which are produced by the division of cambial initials (Larson 1994). Consequently, the term "cambium" is used to refer to cambial initials and "cambial zone" to the region of cambial initials and mother cells (Lachaud *et al.* 1999). Since the initials and mother cells of cambium are distinguished cytologically only by a small difference in length, most published data do not distinguish between the two cell types (*e.g.*, Larson 1994; Savidge 2000b). The terms "cambial cells" and "cambium" will be used hereafter to denote all undifferentiated cells capable of division. Savidge (2000a) argued that every cambial cell could be equally competent and that cambium is maintained in response to basipetally transported auxin in conjunction with physical forces. The differential behaviour of cambial cells could be explained by changes in the micro-environment experienced by the genome within each cell (Savidge 1996).

While the main function of the cambium is cell division and setting out patterns for differentiation, similar to other meristems, several aspects are unique to vascular cambium. Unlike apical meristems, cambium is a complex tissue containing two mor-phologically distinct cell types: axially elongated fusiform cambial cells and somewhat isodiametrical ray cambial cells. These cells give rise to the axial and radial cells of the secondary xylem and phloem. The identity of cambial cells is determined by positional cues rather than by cell lineage, because inter-conversion between fusiform and ray cambial cells is a common phenomenon (*e.g.* Larson 1994; Mellerowicz *et al.* 2001). The balance in the number and distribution between fusiform and ray cambial cells is maintained by anticlinal divisions and by conversion of one kind of cell into the other. Fusiform cambial cells thus give rise to new cambial ray cells through transverse or oblique divisions, while ray cells elongate into fusiform ones through intrusive growth (Lachaud *et al.* 1999). Anticlinal (= radial and pseudotransverse), often also referred to as multiplicative, cell divisions ensure the increase in girth of the cambium (Fig. 1). Transverse and anticlinal divisions ensure the maintenance of cambial integrity, while periclinal (= tangential) divisions (also referred as additive divisions) give rise to new cells of xylem and phloem tissue (Larson 1994; Lachaud *et al.* 1999).

Most divisions of fusiform cambial cells are periclinal, in which new cells are added within a radial file towards either the secondary xylem or the secondary phloem (Catesson 1994). In relation to the functioning of the cambium, two aspects appear to be important for the cambial cell production: 1) the number of dividing cambial cells

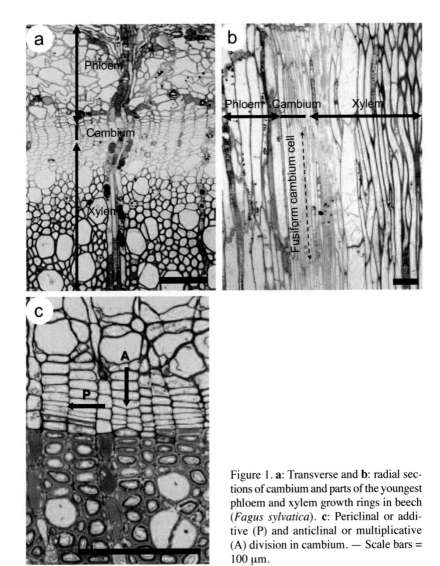

Figure 1. **a**: Transverse and **b**: radial sections of cambium and parts of the youngest
phloem and xylem growth rings in beech
(*Fagus sylvatica*). **c**: Periclinal or additive (P) and anticlinal or multiplicative
(A) division in cambium. — Scale bars =
100 μm.

and how fast the newly formed derivatives are released from the meristematic regio
and 2) the duration of the cell cycle. Both aspects may be individually targeted to
maximise the rate of cell production on the xylem and phloem sides (Uggla *et al.* 1998;
Mellerowicz *et al.* 2001).

Wood and phloem formation are not predetermined processes but are very plastic
expressions of interactions between genotype and the environment (Savidge 2000b) and
require positional information that coordinates the radial pattern of the developmental
zones (Uggla *et al.* 1998). Both phloem and xylem are complex tissues, each containing more than one cell type, so cambial derivatives pass through successive stages
of differentiation during the development of phloem or xylem (Savidge 2000b). Cell

differentiation thus involves four major steps: cell expansion, followed by the ordered deposition of a thick multi-layered secondary cell wall and, in the case of sclerenchyma and tracheary elements, also cell wall lignification and cell death. Cambial cell production is normally more active on the xylem side, explaining the considerable disproportion existing between phloem and xylem tissue (Plomion *et al.* 2001). Fromm (2013) reported that most tree species have xylem to phloem ratios of between 4:1 and 10:1.

SEASONAL CHANGES IN CAMBIAL CELLS

In trees of temperate and cold climatic regions, cambial activity is seasonal and depends on a complex of interactions among intrinsic and extrinsic factors (Savidge 1996; Evert 2006). The dormant period starts immediately after the cessation of meristematic activity and lasts until the resumption of cell divisions (Lachaud *et al.* 1999). Winter dormancy is usually divided into two periods: 1) the resting period or physiological dormancy, which is driven endogenously and 2) quiescence or environmental dormancy, driven by environmental factors (Riding & Little 1984). During the first 2 to 4 weeks of dormancy, the cambium is unable to produce new cells, even when conditions are favourable (resting period). The resting period terminates when the cambium gradually regains the ability to produce new xylem and phloem cells due to favourable environmental conditions (quiescent stage of dormancy).

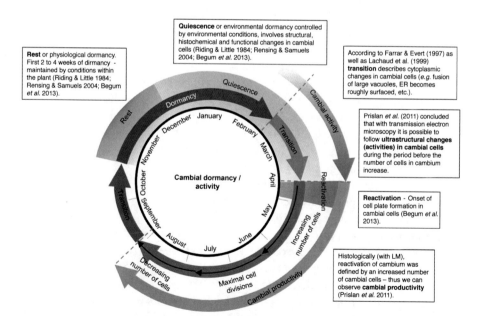

Figure 2. Schematic presentation of most commonly used terminology (proposed by various authors) in terms of cambial dormancy and activity, for the example of *Fagus sylvatica* trees growing at 400 m a.s.l. in Slovenia.

The transition from resting to quiescence involves structural, histochemical and functional changes in cambial cells (Lachaud *et al.* 1999; Begum *et al.* 2013). In addition, this transition between the two phases differs significantly among species (Farrar & Evert 1997b; Lachaud *et al.* 1999). Figure 2 shows the most commonly used terminology in relation to the dormancy and activity of the cambium on the example of *Fagus sylvatica*.

The ultrastructure of cambium cells differs significantly in its active and dormant state and can be seen in the different organisation, distribution, number and shape of the organelles (Farrar & Evert 1997b; Lachaud *et al.* 1999; Rensing & Samuels 2004) (Fig. 3). Most obvious are the differences in shape and size of the vacuoles, which are small, round, numerous and within a dense cytoplasm in dormant cells (Farrar & Evert 1997a; Rensing & Samuels 2004; Frankenstein *et al.* 2005; Prislan *et al.* 2011). In general, the first divisions in spring occur at the end of a 1 to 4 week period characterised by changes in the vacuolar system following the resumption of cyclosis, *i.e.*, the elongation of small vacuoles and their progressive fusion into one or two large vacuoles. Dividing cambial cells contain, among other elements, large vacuoles, rough ER (endoplasmic reticulum), numerous dictyosomes, which produce vesicles and lack storage products such as lipid droplets. The transition from activity to dormancy involves processes whereby large vacuoles fragment into a number of smaller ones, which intersperse throughout the cytoplasm. Rough ER is replaced by smooth ER and an accumulation of storage products takes place, the nature of which depends on the species (Fig. 3). Dictyosomes become fewer and mainly inactive (Farrar & Evert 1997a,b; Lachaud *et al.* 1999; Rensing & Samuels 2004; Prislan *et al.* 2011).

Several authors have reported that lipid droplets are present only in dormant cambium (Robards & Kidwai 1969; Rao & Dave 1983; Farrar & Evert 1997b; Prislan *et al.* 2011). Amyloplasts (starch-containing plastids), however, are numerous in the active state and absent or rare in dormant cambium, as observed in different species (Itoh 1971; Farrar & Evert 1997b; Begum *et al.* 2010a; Prislan *et al.* 2011). Begum *et al.* (2010a) pointed out that storage materials (*e.g.*, lipid droplets and starch-containing plastids) are important for the dormancy and reactivation of cambium. During cambial dormancy, levels of starch might be low as a consequence of the breakdown of starch that is associated with the generation of energy for the development of cold hardiness (Pomeroy & Siminovitch 1971; Timell 1986). Lipid droplets might be utilised as sources of energy for cell division and the biosynthesis of new cell wall material in the cambium (Begum *et al.* 2010a).

Seasonal changes in the ultrastructure of cambial cells may depend on the tree species and sites. Dictyosomes, for example, were found to be active in the dormant period in *Aesculus hippocastanum* (Barnett 1992) and *Pinus strobus* (Srivastava & O'Brien 1966). The appearance of different types of ER (tubular, vesicular or cisternal) in the cambium can also differ between active and dormant states (*e.g.*, Rao & Dave 1983; Farrar & Evert 1997b) and among species (Srivastava & O'Brien 1966; Barnett 1992). These differences can be species specific or can be attributed to differences in growth conditions, different methodological approaches etc.

Seasonal changes of cambial cells are mainly examined in transverse sections; however, ultrastructural features during the cytokinesis of fusiform cambial cells must also be studied in the radial plane. Bailey (1919), for example, described in detail the process of cell divisions in *Pinus strobus* based on observation with light microscope. Although periclinal divisions are most common in cambium, little is known about their ultrastructure (Rensing *et al*. 2002).

In contrast to cells in primary meristems, the length of cambial cells can be 500 times larger than their diameter, so the course of divisions along the axis is slightly different (Rensing *et al*. 2002; Samuels *et al*. 2006). During cambial cell cytokinesis, mitotic spindles separate chromosomes across the radial width/dimension of the cells. However, during cell plate growth, the formed phragmoplast divides axially into two parts. Both parts are surrounded by cytoplasm, representing so-called boluses, which migrate in opposite directions along the extended axis of the cell and form the new cell plate (Rensing *et al*. 2002).

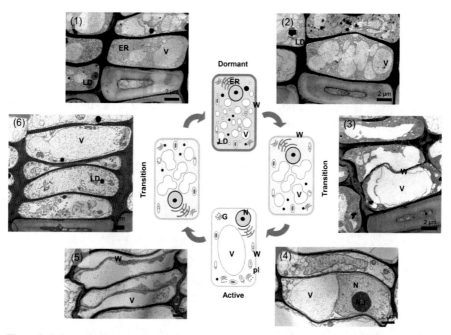

Figure 3. Schematic diagram of cytoplasmic changes occurring in cambial cells of temperate trees during a seasonal cycle. Actively dividing cell in spring or summer, with a large vacuole (V). Transition to rest in autumn; fragmentation of the vacuole (V) and thickening of the cell walls begins (W). Dormant cell in winter; numerous vacuoles become globular during the cessation of cyclosis; mostly smooth endoplasmic reticulum (ER), Golgi apparatus (G) with few secretory vesicles, numerous lipid droplets (LD). Transition to activity in late winter-early spring, showing elongation and fusion of vacuoles following the resumption of cyclosis, rough endoplasmic reticulum, active Golgi apparatus. Nucleus (N) with nucleolus (Nu), plasmodesma (pl). – (1–6) Micrographs showing transverse sections of *Fagus sylvatica* cambial cells in different stages of activity; (1, 2) dormant cells, (3) transition from dormant to active state, (4, 5) active cells and (6) transition from active to dormant state.

Relatively little is known about the structural changes of the cambial cell wall during the transition from the dormant to the active state. In the majority of tree species, the walls of dormant cambial cells are thicker than those in active cells (Larson 1994). Chen *et al.* (2010) found that dormant cambial cell walls of *Populus tomentosa* displayed a multi-layered structure, denser fibril network, smaller pore size and fewer crosslinks between microfibrils than active cambial cell walls. Chaffey *et al.* (1998) suggested, based on observations of cambial seasonal changes in *Aesculus hippocastanum*, that cell-wall thickening at the onset of cambial dormancy should be considered to be secondary thickening and that selective lysis of this secondary wall layer during cambial reactivation restores the thinner, primary wall around active cambial cells. A certain seasonal variability in the frequency of plasmodesmata was also observed in *Populus nigra* (Fuchs *et al.* 2010).

OBSERVATION OF CAMBIAL STRUCTURE WITH LIGHT (LM) AND TRANSMISSION ELECTRON MICROSCOPY (TEM)

Processing plant tissues for TEM can be divided into six major steps: (I) specimen acquisition from a living tree, (II) trimming of the specimen, (III) fixation, (IV) dehydration, (V) infiltration and (VI) embedding (Bozzola & Russell 1999). Standard techniques for wood formation studies are clearly outlined in "Wood formation in trees – cell and molecular biology techniques", edited by Chaffey (2002); from light microscopy to advanced electron microscopy techniques. Recent advances in methodologies for studying the structure of cambium and developing xylem tissue are especially pronounced in the field of tissue fixation; Rensing *et al.* (2002) precisely presented differences in the observation of cambial cells when using conventional chemical fixation or cryofixation and substitution. We demonstrated that combining LM and conventional TEM can provide detailed information on cambial phenology and seasonal ultra-structural changes in cambial cells of *Fagus sylvatica* growing at forest sites with different weather conditions (Prislan *et al.* 2011).

Sample collection

In wood/phloem formation studies at cellular and ultrastructural levels, tissues containing phloem, cambium and outer xylem are collected from living trees. The time intervals of samplings should be relatively short, *i.e.*, one to two weeks, and depend on the goal of the study. For observations of the seasonal dynamics of different phases of wood formation, small micro-cores are usually collected with tools causing minor damage on tree stems. A Trephor tool has recently become widely used for sampling (Rossi *et al.* 2006) but an increment puncher (Forster *et al.* 2000) and injection needles (Jyske *et al.* 2011) are also used. Due to the small size of wounds caused by these tools, repeated sampling on the same tree can be performed in more than one growing season, without affecting the vitality of the tree. However, micro-cores (*e.g.*, diameter c. 2 mm and length c. 10 mm) are difficult to handle and they can be easily damaged or the tissue can be affected. They are therefore not suitable for ultrastructural observations, thus sampling larger blocks of intact tissues using a chisel and knife, as described by Uggla and Sundberg (2002) or Gričar *et al.* (2007b), should be preferred.

Sample preparation and observation

For light microscopic (LM) observations, microcores are first fixed in a solution, such as formalin-ethanol-acetic acid (FEA), dehydrated in a graded series of ethanol and clearing reagent (*e.g.* D-limonene) and infiltrated, as well as embedded in paraffin as described by Rossi *et al.*(2006). Various embedding media, such as glycolmethacrylate (Oberhuber & Gruber 2010) or polyethyleneglycol (Liang *et al.* 2009), are also used (Table 1). Cross sections of 8 to 10 µm thickness are prepared with a rotary microtome and then stained with safranin and astra blue (*e.g.*, Van der Werf *et al.* 2007; Gričar *et al.* 2007a), or cresyl violet acetate (*e.g.*, Antonova & Stasova 1993; Deslauriers *et al.* 2003). For preparation of permanent sections, embedding media such as Euparal (*e.g.*, Gričar *et al.* 2005) or Canada balsam (*e.g.*, Moser *et al.* 2010) are used. Using LM, cambial phenology (onset and cessation of cambial cell production), phases of

Table 1. Overview of sample processing procedures for different tissues and microscopy techniques.

	Light microscopy (LM)	Transmission electron microscopy (TEM)
Use in wood formation studies	Cambial phenology, xylem/phloem differentiation, seasonal dynamics of growth ring formation.	Seasonal changes in cambial ultrastructure (structure of living tissues should be preserved).
Fixation	- Formalin-ethanol-acetic acid solution (FEA) (Gričar *et al.* 2007b); - Ethanol, propionic acid and formaldehyde solution (Oberhuber & Gruber 2010); - Water and ethanol solution (Lupi *et al.* 2010).	*Primary fixative:* - Mixture of 5% glutaraldehyde, 8% paraformaldehyde and 0.3 M cacodylate buffer) (Farrar & Evert 1997b; Frankenstein *et al.* 2005); - Mixture of 2.5% glutaraldehyde in 0.05 M phosphate buffer (pH 6.8) (Rensing & Samuels 2004). *Secondary fixative:* (2% aqueous osmium tetroxide solution).
Dehydration	Graded series of ethanol. *Clearing reagent:* - D-limonene (Gričar *et al.* 2007a); - Histosol (Lupi *et al.* 2010).	Acetone.
Infiltration / embedding	Embedding media: - Paraffin (Rossi *et al.* 2006); - Glycolmethacrylate (Oberhuber & Gruber 2010); - Polyethyleneglycol (Liang *et al.* 2009).	Epoxy resin (Spurr 1969).
Cutting	Rotary microtome (sections 8–12 µm).	Ultra microtome (sections 90–100 nm).
Staining	- Safranin and astrablue (Gričar *et al.* 2007b); - Cresyl violet acetate (Deslauriers *et al.* (2003).	Uranyl acetate and lead citrate.

differentiation of xylem / phloem cells and seasonal dynamics of xylem / phloem growth ring formation can be observed (*e.g.*, Schmitt *et al.* 2000; Gričar 2007; Seo *et al.* 2008; Rossi *et al.* 2011; Michelot *et al.* 2012) (Table 1).

Larger blocks of intact tissue are commonly collected for ultrastructural observations by transmission electron microscopy (TEM), because of easier manipulation and to prevent deformation of the tissue and the occurrence of artefacts. Afterwards, sample size is reduced to less than 2 mm in thickness in order to ensure adequate fixation (Bozzola & Russell 1999). Samples for observation of ultrastructural seasonal changes in cambial cells are fixed for one day in a mixture of 5 % glutaraldehyde, 8 % paraformaldehyde and 0.3 M cacodylate buffer. They are then washed in 0.1 M cacodylate buffer (pH 7.3) and post-fixed for one additional day in a 2 % aqueous osmium tetroxide solution. They are again washed in 0.1 M cacodylate buffer (pH 7.3), dehydrated through a graded series of acetone and finally embedded in Spurr's (1969) epoxy resin. Phosphate buffer is often used instead of cacodylate buffer (Rensing & Samuels 2004) (Table 1). Ultrathin transverse sections (90–100 nm) of cambium are then prepared. Sections are placed on copper grids and stained with uranyl acetate and lead citrate and examined with a TEM at an accelerating voltage of 80 or 100 kV (*e.g.*, Farrar & Evert 1997b; Frankenstein *et al.* 2005; Prislan *et al.* 2011). Seasonal changes in the cytoplasm of cambial cells can be observed by TEM (Farrar & Evert 1997b; Rensing & Samuels 2004), as well as changes in the architecture of cambial cell walls (Chen *et al.* 2010).

ENVIRONMENTAL REGULATION OF CAMBIAL ACTIVITY EXEMPLIFIED BY *FAGUS SYLVATICA*

We illustrate this section with our own examination of cambial phenology in *Fagus sylvatica* at two sites in Slovenia, central Europe, with different elevations and weather regimes. The low elevation forest site (400 m a.s.l.) has a mean annual temperature (MAT) of 11.3 °C and 1565 mm of annual precipitation and the high elevation site (1200 m a.s.l.) has 6.6 °C MAT and slightly higher precipitation (Prislan *et al.* 2011; 2013).

The two sites were carefully selected based on previous investigations in beech and climatic factors in Slovenia involving tree-ring variation and climate (Di Filippo *et al.* 2007; Čufar *et al.* 2008), leaf phenology (Čufar *et al.* 2012), climatic situation and trends (De Luis *et al.* 2012), and connection of tree-ring variation, leaf phenology, cambial activity and wood formation and climate (Čufar *et al.* 2008; Prislan *et al.* 2013). Thus, the selected locations are representative for growth of beech at low and high elevations in Slovenia.

Dormant cambium contained 4 to 5 cell layers at both sites, whereas active cambium had a slightly higher number of cell layers at the low elevation site (around 11) than at the high elevation site (around 8) (Prislan *et al.* 2011). At the high elevation site, the onset of cambial cell production occurred one month later (in the middle of May) than at the low elevation site (Prislan *et al.* 2013). Maximal cell productivity was observed at the high elevation site around the summer solstice and at the low elevation site at the beginning of June. Cell production ceased at the beginning of August at the high elevation site and at the end of August at the low elevation site (Fig. 4) (Prislan *et al.* 2011).

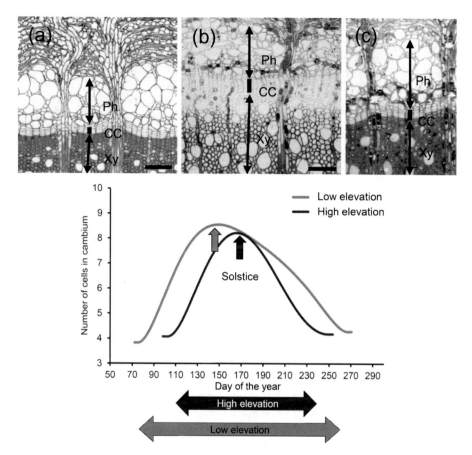

Figure 4. Light micrographs showing the cambium of *Fagus sylvatica* (a) in the dormant state before the onset and (c) after the cessation of cambial activity and (b) active cambium. The graph shows seasonal changes in the number of cells in the cambium at high and low elevation sites (d). Arrowheads indicate the timing of maximal cambial productivity; at the high elevation site this occurred, around the summer solstice, at the low elevation site in the beginning of June. Arrows show the duration of cambial cell production.

Our study demonstrated that the duration of the growing season is inversely proportional to the altitude (in temperate regions); the onset of cell production at a high elevation starts later and ceases earlier (Prislan *et al.* 2013). However, this conclusion is preliminary and for its generalisation about the effect of elevation on cambial phenology further studies should be performed with the inclusion of additional sites. The shorter duration of cambial activity at the higher sites can be ascribed to the generally lower air and soil temperatures, as well as a longer period of snow cover, since such conditions are limiting for physiological processes, particularly at the beginning of the vegetation period, as reported for instance by Kirdyanov *et al.* (2003) and Moser *et al.* (2010).

The cambium is mainly susceptible to environmental signals during the active period and archives them in the wood and bark structure (*e.g.*, Frankenstein *et al.* 2005; Gričar & Čufar 2008).

In temperate and cool regions, temperature and photoperiod are important external factors of the initiation of cambial reactivation and xylem differentiation (Li *et al.* 2009; Begum *et al.* 2013). Deslauriers *et al.* (2008) studied cambial phenology in *Pinus leucodermis* at high altitudes in Italy and showed that temperatures in spring are the main factor driving the onset of cell production. Rossi *et al.* (2008) tried to define temperature thresholds for the onset and cessation of cambial cell production for conifers from different latitudes of boreal and temperate regions. The presented mean critical temperatures, which varied by around 8 °C, were significant for the onset of cell production; however, mean critical temperatures for cessation (around 14 °C) were not significant. Seo *et al.* (2011) monitored the intra-annual growth dynamics of pine trees (*Pinus sylvestris*) in northern Scandinavia, with diverging results; at some sites, wood formation was mainly positively correlated with temperature, whereas such a positive correlation was missing at other sites or it was even negative. Furthermore, we also showed that the temperatures prior to the occurrence of relevant cambial phenological phases, together with calculated growing degree days, significantly differed at the low and high elevation beech sites in Slovenia, indicating that phenological events are not in simple relationship with climate, or at least not in agreement with year-to-year variations in weather. The accumulated heat units (growing degree days - GGDs) at the beginning of cambial activity, for instance, were higher in the lowland (around 147 °D) than at higher elevations (around 72 °D) (Prislan *et al.* 2013).

Seasonal changes in the ultrastructure of cambial cells studied along the altitudinal gradient are rare. We showed by TEM that processes in cambial cells that are related to seasonal changes are similar regardless of the elevation; only the timings of individual events differ and are generally delayed at higher elevations at the onset of the growing season.

At the end of the growing season, the sequence of changes in the cambium was just the opposite; the transition from active to dormant state started first in the cambial cells of *Fagus sylvatica* from the higher elevation (Prislan *et al.* 2011).

LM AND TEM, COMPLEMENTARY APPROACHES TO STUDYING CAMBIUM SEASONALITY

This section is also illustrated by our own results obtained by a combination of TEM and LM studies of *Fagus sylvatica* from two elevations in Slovenia. TEM observations revealed that cytoplasmic changes in cambial cells during the seasonal cycle occur much earlier (about one month) than was observed at the cellular level with LM (Prislan *et al.* 2011).

The results are affected by: 1) the media used for tissue fixation and embedding, as well as 2) the thickness of sections and microscope resolution/magnification. A fixative such as FAA (formaldehyde, ethanol and acetic acid) is commonly used in wood formation studies (observing cambial phenology and xylem growth ring formation dynamic), whereby sections are observed with LM. The main disadvantage of FAA is insufficient preservation of the cytoplasm in cambial cells. Reactivation of the cambium at the cellular level is usually histologically defined by an increased number of cambial cells and the occurrence of newly formed xylem and phloem cells in early developmental stages. With LM, it is possible to detect cambial production based on the number of newly formed cells at a certain time interval, as well as differentiation of xylem and phloem cells; however, cytoplasmic changes associated with the seasonal cycle of the cambium cannot be recognised (Prislan *et al.* 2011).

Observations on a sub-cellular level are also possible with LM (Fig. 5), when using primary fixation with glutaraldehyde and phosphate buffer and secondary fixation in osmium tetroxide (usually used for TEM observations), as demonstrated by Bailey (1919) or Begum *et al.* (2010b), who were able to observe new cell plates. Similarly, Rensing and Samuels (2004) were able to observe differences in the arrangement of vacuoles between dormant and active cambial cells.

TEM, with its high resolution and magnification (Goodhew *et al.* 2000), in combination with proper fixation of the cambial tissue, *i.e.*, preservation of cytoplasm, allows the observation of seasonal changes in the distribution and size of cell organelles (Fig. 5). This is particularly important during the transition of cambial cells to active or dormant state, when the number of cell layers is unchanged but ultra-structural changes (in the form of dictyosome activity, changes related to endoplasmic reticulum, etc.) are already occurring.

Consequently, different criteria for cambial reactivation are used with different sample preparation and microscopy techniques (resolutions), as was already stressed by Frankenstein *et al.* (2005). Begum *et al.* (2013), *e.g.*, called the period from late winter to late spring, when new cell plates are formed in the cambium, "cambial reactivation". In our TEM study (Prislan *et al.* 2011), changes in the ultrastructure of cambial cells were observed prior to the formation of new cell walls, and the cambium can thus be considered to be active as well in this earlier stage. When using different fixation, embedding and microscopy techniques and criteria, the established dates of cambial seasonality can vary. However, a combination of different approaches enables the activity and productivity of cambium to be precisely followed.

CONCLUSIONS

Examination of cambial tissue using different microscopic levels revealed that different criteria are used to define the onset of cambial reactivation; the results of different studies are therefore not simply comparable. The terminology should thus be adjusted and standardised to avoid disagreement among different research groups when comparing data, as was already stressed by Frankenstein *et al.* (2005).

Complementary methodologies can assure a better understanding of relations between physiological processes in the cambium and climate with regard to predictions of tree responses to anticipated climate change. The presented facts can help in future cambial studies, particularly when the results of different studies using different techniques must be compared.

ACKNOWLEDGEMENTS

This work was supported by the Slovenian Research Agency, young researchers' program (Peter Prislan) programs P4-0015 and P4-0107, and by the LLP ERASMUS bilateral agreement between the University of Ljubljana and the University of Hamburg. The cooperation among international partners was supported by the COST Action FP1106, STReESS. The authors gratefully acknowledge the help of Professor Jasna Štrus and her team at the Department of Biology, Biotechnical Faculty, University of Ljubljana. We thank Tanja Potsch, Dr. Magda Tušek Žnidarič and Dr. Nada Žnidaršič for their immense help in the laboratory.

REFERENCES

Antonova GF & Stasova VV. 1993. Effects of environmental factors on wood formation in Scots pine stems. Trees 7: 214–219.

Bailey IW. 1919. Phenomena of cell division in the cambium of arborescent gymnosperms and their cytological significance. Proceedings of the National Academy of Sciences of the USA 5: 283–285.

Barnett JR. 1992. Reactivation of the cambium in *Aesculus hippocastanum* L. – a transmission electron microscopy study. Ann. Bot. 70: 169–177.

Begum S, Nakaba S, Oribe Y, Kubo T & Funada R. 2010a. Changes in the localization and levels of starch and lipids in cambium and phloem during cambial reactivation by artificial heating of main stems of *Cryptomeria japonica* trees. Ann. Bot. 106: 885–895.

Begum S, Nakaba S, Oribe Y, Kubo T & Funada R. 2010b. Cambial sensitivity to rising temperatures by natural condition and artificial heating from late winter to early spring in the evergreen conifer *Cryptomeria japonica*. Trees 24: 43–52.

Begum S, Nakaba S, Yamagishi Y, Oribe Y & Funada R. 2013. Regulation of cambial activity in relation to environmental conditions: understanding the role of temperature in wood formation of trees. Physiol. Plantarum 147: 46–54.

Bozzola JJ & Russell LD. 1999. Electron microscopy: principles and techniques for biologists. Jones and Bartlett Publishers, Boston.

Callado CH, Roig FA, Tomazello-Filho M & Barrow CF. 2013. Cambial growth periodicity studies of South American woody species – A review. IAWA J. 34: 213–230.

Catesson A-M. 1994. Cambial ultrastructure and biochemistry: changes in relation to vascular tissue differentiation and the seasonal cycle. Int. J. Plant Sci. 155: 251–261.

Chaffey NJ. 2002. Wood formation in trees: cell and molecular biology techniques. Taylor & Francis, London, New York.

Chaffey NJ, Barlow PW & Barnett JR. 1998. A seasonal cycle of cell wall structure is accompanied by a cyclical rearrangement of cortical microtubules in fusiform cambial cells within taproots of *Aesculus hippocastanum* (Hippocastanaceae). New Phytol. 139: 623–635.

Chen HM, Han JJ, Cui KM & He XQ. 2010. Modification of cambial cell wall architecture during cambium periodicity in *Populus tomentosa* Carr. Trees 24: 533–540.

Costa MS, de Vasconcellos TJ, Barros CF & Callado CH. 2013. Does growth rhythm of a widespread species change in distinct growth sites? IAWA J. 34: 498–509.

Čufar K, De Luis M, Saz M, Črepinšek Z & Kajfež-Bogataj L. 2012. Temporal shifts in leaf phenology of beech (*Fagus sylvatica*) depend on elevation. Trees 26: 1091–1100.

Čufar K, Prislan P, De Luis M & Gričar J. 2008. Tree-ring variation, wood formation and phenology of beech (*Fagus sylvatica*) from a representative site in Slovenia, SE Central Europe. Trees 22: 749–758.

De Luis M, Čufar K, Saz M, Longares L, Ceglar A & Kajfež-Bogataj L. 2012. Trends in seasonal precipitation and temperature in Slovenia during 1951–2007. Reg. Environ. Change. DOI: 10.1007/s10113-012-0365-7.

Deslauriers A, Morin H & Begin Y. 2003. Cellular phenology of annual ring formation of *Abies balsamea* in the Quebec boreal forest (Canada). Can. J. Forest Res. 33: 190–200.

Deslauriers A, Rossi S, Anfodillo T & Saracino A. 2008. Cambial phenology, wood formation and temperature thresholds in two contrasting years at high altitude in southern Italy. Tree Physiol. 28: 863–871.

Di Filippo A, Biondi F, Čufar K, De Luis M, Grabner M, Maugeri M, Presutti Saba E, Schirone B & Piovesan G. 2007. Bioclimatology of beech (*Fagus sylvatica* L.) in the Eastern Alps: spatial and altitudinal climatic signals identified through a tree-ring network. J. Biogeogr. 34: 1873–1892.

Evert RF. 2006. Esau's Plant anatomy. Meristems, cells, and tissues of the plant body: their structure, function, and development. Wiley-Interscience, Hoboken, New Jersey.

Farrar JJ & Evert RF. 1997a. Ultrastructure of cell division in the fusiform cells of the vascular cambium of *Robinia pseudoacacia*. Trees 11: 203–215.

Farrar JJ & Evert RF. 1997b. Seasonal changes in the ultrastructure of the vascular cambium of *Robinia pseudoacacia*. Trees 11: 191–202.

Forster T, Schweingruber FH & Denneler B. 2000. Increment puncher – A tool for extracting small cores of wood and bark from living trees. IAWA J. 21: 169–180.

Frankenstein C, Eckstein D & Schmitt U. 2005. The onset of cambium activity – A matter of agreement? Dendrochronologia 23: 57–62.

Fromm J. 2013. Xylem development in trees: From cambial divisions to mature wood cells. In: Fromm J (ed.), Cellular aspects of wood formation: 3–39. Springer, Berlin, Heidelberg.

Fuchs M, van Bel AJE & Ehlers K. 2010. Season-associated modifications in symplasmic organization of the cambium in *Populus nigra*. Ann. Bot. 105: 375–387.

Goodhew PJ, Humphreys J & Beanland R. 2000. Electron microscopy and analysis. Taylor and Francis, London, New York.

Gričar J. 2007. Xylo- and phloemogenesis in silver fir (*Abies alba* Mill.) and Norway spruce (*Picea abies* (L.) Karst.). Slovenian Forestry Institute, Ljubljana.

Gričar J & Čufar K. 2008. Seasonal dynamics of phloem and xylem formation in silver fir and Norway spruce as affected by drought. Russ. J. Plant Physiol. 55: 538–543.

Gričar J, Čufar K, Oven P & Schmitt U. 2005. Differentiation of terminal latewood tracheids in silver fir trees during autumn. Ann. Bot. 95: 959–965.

Gričar J, Zupančič M, Čufar K, Koch G, Schmitt U & Oven P. 2006. Effect of local heating and cooling on cambial activity and cell differentiation in the stem of Norway spruce (*Picea abies*). Ann. Bot. 97: 943–951.

Gričar J, Zupančič M, Čufar K & Oven P. 2007a. Wood formation in Norway spruce (*Picea abies*) studied by pinning and intact tissue sampling method. Wood Res. 52: 1–10.

Gričar J, Zupančič M, Čufar K & Oven P. 2007b. Regular cambial activity and xylem and phloem formation in locally heated and cooled stem portions of Norway spruce. Wood Sci. Technol. 41: 463–475.

Itoh T. 1971. On the ultrastructure of dormant and active cambium of conifers. Bull. Wood Res. Inst. Kyoto Univ. 51: 33–45.

Jyske T, Manner M, Mäkinen H, Nöjd P, Peltola H & Repo T. 2011. The effects of artificial soil frost on cambial activity and xylem formation in Norway spruce. Trees 26: 1–15.

Kirdyanov A, Hughes M, Vaganov E, Schweingruber F & Silkin P. 2003. The importance of early summer temperature and date of snow melt for tree growth in the Siberian Subarctic. Trees 17: 61–69.

Lachaud S, Catesson AM & Bonnemain JL. 1999. Structure and functions of the vascular cambium. Life Sci. 322: 633–650.

Larson PR. 1994. The vascular cambium: development and structure. Springer, Berlin.

Li WF, Ding Q, Chen JJ, Cui KM & He XQ. 2009. Induction of PtoCDKB and PtoCYCB transcription by temperature during cambium reactivation in *Populus tomentosa* Carr. J. Exp. Bot. 60: 2621–2630.

Liang EY, Eckstein D & Shao XM. 2009. Seasonal cambial activity of relict Chinese pine at the northern limit of its natural distribution in north China - Exploratory results. IAWA J. 30: 371–378.

Lupi C, Morin H, Deslauriers A & Rossi S. 2010. Xylem phenology and wood production: resolving the chicken-or-egg dilemma. Plant Cell Environ. 33: 1721–1730.

Mauseth JD. 2009. Botany: An introduction to plant biology. Jones & Bartlett Publ., Sudbury.

Mellerowicz EJ, Baucher M, Sundberg B & Boerjan W. 2001. Unravelling cell wall formation in woody dicot stem. Plant Mol. Biol. 47: 239–274.

Michelot A, Simard S, Rathgeber C, Dufrêne E & Damesin C. 2012. Comparing the intra-annual wood formation of three European species (*Fagus sylvatica*, *Quercus petraea* and *Pinus sylvestris*) as related to leaf phenology and non-structural carbohydrate dynamics. Tree Physiol. 32: 1033–1045.

Moser L, Fonti P, Büntgen U, Esper J, Luterbacher J, Franzen J & Frank D. 2010. Timing and duration of European larch growing season along altitudinal gradients in the Swiss Alps. Tree Physiol. 30: 225–233.

Oberhuber W & Gruber A. 2010. Climatic influences on intra-annual stem radial increment of *Pinus sylvestris* (L.) exposed to drought. Trees 24: 887–898.

Plomion C, LeProvost G & Stokes A. 2001. Wood formation in trees. Plant Physiol. 127: 1513–1523.

Pomeroy MK & Siminovitch D. 1971. Seasonal cytological changes in secondary phloem parenchyma cells in *Robinia pseudoacacia* in relation to cold hardiness. Can. J. Bot. 49: 787–795.

Prislan P, Gričar J, De Luis M, Smith KT & Čufar K. 2013. Phenological variation in xylem and phloem formation in *Fagus sylvatica* from two contrasting sites. Agr. Forest Meteorol. 180: 142–151.

Prislan P, Schmitt U, Koch G, Gričar J & Čufar K. 2011. Seasonal ultrastructural changes in the cambial zone of beech (*Fagus sylvatica*) grown at two different altitudes. IAWA J. 32: 443–459.

Rao KS & Dave YS. 1983. Ultrastructure of active and dormant cambial cells in teak (*Tectona grandis* L.f.). New Phytol. 93: 447–456.

Rathgeber CBK, Rossi S & Bontemps J-D. 2011. Cambial activity related to tree size in a mature silver-fir plantation. Ann. Bot. 108: 429–438.

Rensing KH & Samuels AL. 2004. Cellular changes associated with rest and quiescence in winter-dormant vascular cambium of *Pinus contorta*. Trees 18: 373–380.

Rensing KH, Samuels AL & Savidge RA. 2002. Ultrastructure of vascular cambial cell cytokinesis in pine seedlings preserved by cryofixation and substitution. Protoplasma 220: 39–49.

Riding RT & Little CHA. 1984. Anatomy and histochemistry of *Abies balsamea* cambial zone cells during the onset and breaking of dormancy. Can. J. Bot. 62: 2570–2579.

Risopatron JPM, Sun YQ & Jones BJ. 2010. The vascular cambium: molecular control of cellular structure. Protoplasma 247: 145–161.

Robards AW & Kidwai P. 1969. A comparative study of the ultrastructure of resting and active cambium of *Salix fragilis* L. Planta 84: 239–224.

Rossi S, Anfodillo T & Menardi R. 2006. Trephor: a new tool for sampling microcores from tree stems. IAWA J. 27: 89–97.

Rossi S, Deslauriers A, Gričar J, Seo JW, Rathgeber CWG, Anfodillo T, Morin H, Levanič T, Oven P & Jalkanen R. 2008. Critical temperatures for xylogenesis in conifers of cold climates. Global Ecol. Biogeogr. 17: 696–707.

Rossi S, Morin H & Deslauriers A. 2011. Multi-scale influence of snowmelt on xylogenesis of black spruce. Arctic, Antarctic, and Alpine Research 43: 457–464.

Samuels AL, Kaneda M & Rensing KH. 2006. The cell biology of wood formation: from cambial divisions to mature secondary xylem. Can. J. Bot. 84: 631–639.

Savidge RA. 1996. Xylogenesis, genetic and environmental regulation – a review. IAWA J. 17: 269–310.

Savidge RA. 2000a. Intrinsic regulation of cambial growth. J. Plant Growth Regul. 20: 52–77.

Savidge RA. 2000b. Biochemistry of seasonal cambial growth and wood formation – an overview of the challenges. In: Savidge RA, Barnett JR, Napier R &Biggs A (eds.), Cell and molecular biology of wood formation: 1–30. BIOS Scientific Publishers Ltd, Oxford, UK.

Schmitt U, Möller R & Eckstein D. 2000. Seasonal wood formation dynamics of beech (*Fagus sylvatica* L.) and black locust (*Robinia pseudoacacia* L.) as determined by the "pinning" technique. J. Appl. Bot. 74: 10–16.

Seo JW, Eckstein D, Jalkanen R, Rickebusch S & Schmitt U. 2008. Estimating the onset of cambial activity in Scots pine in northern Finland by means of the heat-sum approach. Tree Physiol. 28: 105–112.

Seo JW, Eckstein D, Jalkanen R & Schmitt U. 2011. Climatic control of intra- and interannual wood-formation dynamics of Scots pine in northern Finland. Environ. Exp. Bot. 72: 422–431.

Spurr AR. 1969. A low viscosity embedding medium for electron microscopy. J. Ultrastruct. Res. 26: 31–43.

Srivastava LM & O'Brien TP. 1966. On the ultrastructure of cambium and its vascular derivatives. I. Cambium of *Pinus strobus* L. Protoplasma 61: 257–276.

Timell TE. 1986. Formation of compression wood. In: (ed.), Compression wood in gymnosperms 1: 623–636. Springer-Verlag, Berlin.

Uggla C, Mellerowicz EJ & Sundberg B. 1998. Indole-3-acetic acid controls cambial growth in Scots pine by positional signaling. Plant Physiol. 117: 113–121.

Uggla C & Sundberg B. 2002. Sampling of cambial region tissues for high resolution analysis. In: Chaffey N (ed.), Wood formation in trees. Cell and molecular biology techniques: 215–228. Taylor & Francis, London, New York.

Ursache R, Nieminen K & Helariutta Y. 2013. Genetic and hormonal regulation of cambial development. Physiol. Plantarum 147: 36–45.

Van der Werf GW, Sass-Klaassen U & Mohren GMJ. 2007. The impact of the 2003 summer drought on the intra-annual growth pattern of beech (*Fagus sylvatica* L.) and oak (*Quercus robur* L.) on a dry site in the Netherlands. Dendrochronologia 25: 103–112.

Wodzicki TJ. 2001. Natural factors affecting wood structure. Wood Sci. Tech. 35: 5–26.

Zimmermann HM & Brown CL. 1971. Trees structure and function. Springer, Berlin, Heidelberg, New York.

Accepted: 27 August 2013

IAWA Journal 34 (4), 2013: 408–424

BRILL

VISUALIZING WOOD ANATOMY IN THREE DIMENSIONS
WITH HIGH-RESOLUTION X-RAY MICRO-TOMOGRAPHY (MCT)
– A REVIEW –

Craig R. Brodersen

Horticultural Sciences Department, Citrus Research & Education Center, University of Florida,
700 Experiment Station Road, Lake Alfred, FL 33850, U.S.A.
E-mail: brodersen@ufl.edu

ABSTRACT

High-resolution X-ray micro-tomography (μCT) has emerged as one of the most promising new tools available to wood anatomists to study the three-dimensional organization of xylem networks. This non-destructive method faithfully reproduces the spatial relationships between the different cell types and allows the user to explore wood anatomy in new and innovative ways. With μCT imaging, the sample can be visualized in any plane and is not limited to a single section or exposed plane. Conventional CT software aids in the visualization of wood structures, and newly developed custom software can be used to rapidly automate the data extraction process, thereby accelerating the rate at which samples can be analyzed for research. In this review the origins of xylem reconstructions using traditional methods are discussed, as well as the current applications of μCT in plant biology and an overview of pertinent technical considerations associated with this technique. μCT imaging offers a new perspective on wood anatomy and highlights the importance of the relationships between wood structure and function.

Keywords: Synchrotron, 3D, tomography, wood anatomy, visualization, μCT.

INTRODUCTION

Over the past 10–15 years high resolution X-ray micro-computed tomography (μCT) has seen a surge in popularity as a tool for producing three-dimensional (3D) visualizations of plant tissue. This non-destructive method allows wood anatomists to repeatedly section a single block of wood from any perspective, selectively isolate a region of interest, and then repeat the same process *ad infinitum* throughout the sample with each iteration of this process digitally preserved. μCT technology has progressed to the point where it now rivals low magnification scanning electron microscopy (SEM) in its ability to resolve fine details. However, SEM can only visualize an exposed surface while μCT can be used to probe the entire sample and virtually expose a plane in any orientation. As this technology continues to mature, image resolution, quality, and acquisition time will improve and μCT has the potential to serve as an important tool in the analysis of wood structure and function. Increased interest in this new technique

© International Association of Wood Anatomists, 2013
Published by Koninklijke Brill NV, Leiden

DOI 10.1163/22941932-00000033

is indicative of the recognized importance of xylem network connectivity by wood anatomists and the inherent difficulty of visualizing the spatial organization of xylem networks in three dimensions.

Studying the spatial organization of xylem vessels is challenging not only because of the scale at which the networks exist, but also because of the tools traditionally available to visualize such networks. Light microscopy is the most common visualization method, which applies a two-dimensional tool to a three-dimensional problem. In many species, vessels do not follow a straight course through the wood, but instead "drift" laterally around the stem, often resulting in spiral grain (Zimmermann & Brown 1971; André 2005). As a consequence, aligning a microtome to cut a radial or tangential plane through multiple vessels to track their ascent through a large sample of wood is exceedingly difficult, and a single transverse section reveals little about the axial course of a vessel or vessel group. Serial sectioning, therefore, has been the most popular method for reconstructing xylem networks (*e.g.* Burggraaf 1972; Bosshard & Kučera 1973), continues to be used regularly (Fujii *et al.* 2001; Kitin *et al.* 2004), and is one of the fundamental components of confocal microscopy (Kitin *et al.* 2003). From those sections the path or course of individual xylem conduits can be tracked through a length of plant tissue such that the connections between vessels can be determined.

This exercise, while time consuming and at times tedious, is highly recommended for anyone interested in studying the spatial organization of the xylem, as it is one of the easiest ways to develop a three-dimensional understanding of the spatial relationships between different plant tissues. The serial sections and reconstructed xylem network reveal that vessels are not merely straight, vertical pipes, but dynamic structures that move or "drift" radially or tangentially around the stem. Serial sectioning also allows the viewer to study the elegant solutions plants have developed for the distribution of water through the bifurcation of vascular bundles that lead to leaf traces (McCulloh *et al.* 2003) or the rich diversity in stelar organization (Beck *et al.* 1982; Pittermann *et al.* 2013), often revealing relationships between structure and function of the xylem network that are otherwise obscured by the complexity of plant tissue (Brodersen *et al.* 2012).

The development of the optical shuttle method was one of the key advancements in the reconstruction of xylem networks using serial sections (Zimmermann & Tomlinson 1966). The system developed by Zimmermann and Tomlinson (1966) featured a 16 mm film camera mounted to a microscope focused on the exposed transverse face of a stem sample mounted in a microtome. Following the removal of each transverse section a photograph was taken, and the serial images were then assembled into a film. Each frame could then be viewed sequentially in forward or reverse to scroll axially along the stem, where the z-axis of the stem was translated into time (*e.g.* Zimmermann & Brown 1971; Zimmermann & Tomlinson 1968). This method significantly decreased the time necessary to process a stem sample and preserved the sections as photographic images that could be reconstructed at a later time. However, reconstructing the network was performed manually by projecting the film onto tracing paper where the position of individual vessels could be tracked through a length of stem (Zimmermann & Brown 1971; Fig. 1). The resulting reconstructions revealed the complex and convoluted nature

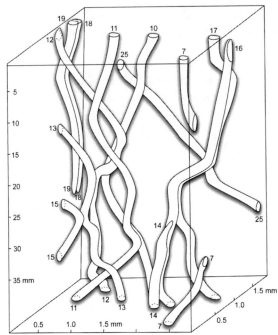

Figure 1. Partial reconstruction of the vessel network of *Cedrela fissilis* redrawn from Zimmermann and Brown (1971). Serial light micrographs were used to plot the course of vessels through a block of wood visualized using the optical shuttle method. Each vessel is numbered at the end where it enters or leaves the block of wood. The y-axis was foreshortened ten times.

of the vessel networks in many species. The method has been updated to couple serial sectioning with digital micro-photography, yielding excellent results (Kitin *et al.* 2004; André 2005; Huggett & Tomlinson 2010; Wu *et al.* 2011).

The original films are still of use today because of the high quality of the serial sections, many of which have been digitized. As an example, one such film (Zimmermann 1971) was loaded into freely available software (Quicktime 7.0.1, Apple Computer Inc.; FIJI image processing software (a Java-based distribution of ImageJ)), and the individual frames of the film were extracted as an image sequence. Because of the known frame rate in the digital version, the original frame rate of the film version, and the section thickness, the image sequence was easily transformed into a 3D volume rendering using commercially available software (Avizo 7.0, VSG) (Fig. 2). The whole block of wood can be reconstructed from the image sequence (Fig. 2b), and individual vessels can be selected, reproduced as volume renderings, and viewed from a variety of angles to study the course of the vessels through the wood (Fig. 2c, d). Much like the original films, the axial scale can be compressed (Fig. 1, 2b–d), or displayed to show the true scale of the sample (Fig. 2e). While this example shows that films produced some 40 years ago can be quickly reconstructed using modern computer software, it should be noted that this method is easily adaptable to serial sections created today. While the optical shuttle equipment makes the process much faster, traditional serial sections can be used with the same software to make reconstructions of xylem networks. Currently, μCT systems often have limited access, and digital serial sections created using light microscopy could be a simple and inexpensive alternative if imaging in a single plane is sufficient.

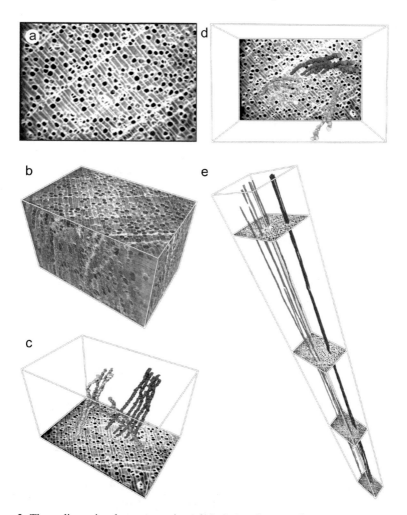

Figure 2. Three-dimensional reconstruction of *Cedrela odorata* xylem vessels using visualization software. Transverse serial images of the wood (a) were extracted from the film produced by Zimmermann and Brown (1971) by isolating individual frames in the film and then loading them into 3D visualization software. The block of wood could then be visualized as a whole (b), foreshortened in the y-axis as originally presented by Zimmermann and Brown (1971), as selected vessel groups (c, d), or fully expanded in the y-axis (e). The software readily allows the user to visualize the network from different perspectives (c, d) to track the course of the vessels through the wood. The long edge of the wood block is X μm, and scale varies with perspective in b–e. In (e) the length of the block is X μm.

TECHNICAL CONSIDERATIONS FOR VISUALIZING WOOD ANATOMY WITH MCT IMAGING

Plant research utilizing μCT has proliferated over the past decade and these studies were facilitated by the presence of lab-based μCT systems in university imaging facilities and the development of synchrotron-based microtomography instruments at a limited number of facilities around the world (*e.g.* the Lawrence Berkeley National Laboratory Advanced Light Source (USA), Swiss Light Source (Switzerland), Australian Synchrotron, etc.).

Lab-based systems now offer comparable image resolution compared to synchrotron instruments, but the primary advantage of synchrotron μCT is the high flux that allows for shorter exposure times, and therefore shorter overall scan times, and a wider range of available X-ray energies. While μCT is not the only non-destructive method for visualizing plant tissue (Bucur 2003), the advantages of using μCT for studying wood anatomy is becoming clear. Magnetic resonance imaging is an alternative to μCT for 3D imaging, and Oven *et al.* (2011) have shown the utility of this tool for studying plant tissue at a lower resolution than μCT, but free of any issues related to X-ray exposure.

The μCT instruments used for studying plants are based on the same principles as medical CT systems. Throughout the literature the method has been given many names and the field has yet to settle on a specific acronym (*e.g.* micro computed tomography (μCT), high-resolution computed tomography (HRCT), high-resolution X-ray computed tomography (HRXCT), X-ray micro computed tomography (XMCT), etc.); however, they all refer to the same method with slight variations depending on the X-ray source and facility configuration. Briefly, X-rays aimed at a sample are attenuated based on the absorption properties of the sample's constituents. Opposite the X-ray source is a scintillator that converts X-rays into visible light, which is then directed with a series of lenses and mirrors to a CCD camera that captures a single projection image. The current CCD camera utilized by the Advanced Light Source in Berkeley, CA has a 4008 × 2672 pixel array, and at the 4.5 μm resolution the field of view is 18 × 12 mm. The sample is then rotated in small increments (*e.g.* 0.125°) over 180°, with a projection image taken after each angular increment. These projection images are then normalized for image intensity, background corrected, and then "reconstructed" into a set of digital serial images composed of voxels (volumetric pixel elements) instead of pixels. Each voxel is assigned an x, y, and z coordinate and an intensity value corresponding to X-ray attenuation for that point in three-dimensional space. The result is a stack of digital images not unlike a set of serial light micrographs. Contrast in μCT imaging is achieved through the natural attenuation of X-rays by the sample. Regions within the sample that absorb fewer X-rays appear as dark voxels, and dense areas appear white or light gray voxels. Cellulose and other carbon-based compounds readily absorb X-rays, and the difference in X-ray attenuation between plant tissue and the surrounding air provides excellent contrast. Contrast agents can be injected into xylem vessels (*e.g.* KI solutions or silicone resin, see Brodersen *et al.* 2011), but most conventional (*i.e.* medical) contrast agents are incompatible with live plant tissue or readily diffuse through the porous cell walls of dry plant tissue. Many image visualization software packages are available for visualizing μCT datasets, including both commercial (Avizo (VSG, Inc., Burlington, Massachusetts, USA); VGStudioMax (Volume Graphics GmbH, Heidelberg, Germany), etc.) and free options (Fiji (www.fiji.sc/Fiji); Drishti (anusf.anu.edu.au/Vizlab/drishti)). In addition, several wood anatomy research groups have developed custom software packages to automate the analysis of wood μCT images (*e.g.* Steppe *et al.* 2004; Brodersen *et al.* 2011).

Dried plant tissue is particularly amenable to μCT imaging because of the relatively large difference in X-ray attenuation between plant tissue and air inside the vessel or tracheid lumen. Small diameter cells (*e.g.* < 20 μm) are less easily visualized at lower resolution (*e.g.* > 5 μm) and benefit from high resolution scans. In live plant tissue or excised fresh tissue that remains hydrated, the air-filled xylem conduits are easily distinguished, while the surrounding water-filled fibers and parenchyma are difficult to visualize. However, phase contrast μCT can help to enhance the outline of the cell walls, but contrast in the voxel intensity between the cellular water and the cell walls is often insufficient to separate the two substances during image segmentation (Brodersen *et al.* 2011). Recently, Blonder *et al.* (2012) used μCT (lab-based and synchrotron) to show leaf venation could be imaged using this technique, providing an alternative to traditional methods for clearing leaves.

As noted above, the resolving power of μCT instruments continues to improve, and many facilities allow imaging at three to four different magnifications. However, increasing the resolution decreases the field of view, not unlike traditional light microscopy optics, and resolution should be selected based on the requirements of the investigation. For example, a xylem network composed of vessels approximately 20 μm in diameter scanned with a 5 μm resolution would yield an image with four voxels that span the vessel lumen. In most μCT systems some image noise is inevitable, and low signal:noise can lead to images with low contrast between the vessel lumen and the surrounding plant tissue. With such a low sampling of voxels inside the lumen visualizing the important details (*e.g.* connections, pitting, vessel endings) can be difficult. One solution is to increase resolution at the expense of a smaller field of view. In our example, decreasing the voxel size to 2 μm or smaller effectively doubles the number of voxels representing the vessel lumen. Most μCT systems allow for "tiling", where the sample can be shifted in the X-ray path such that subsequent scans can seamlessly capture an adjoining area of the sample with only 5–10 μm of overlap required between scans for registration. Using image-processing software these tiles can be combined to create a continuous dataset. This technique can be useful for scanning at high resolution to study xylem organization outside of the field of view of a single scan, or allow the user to scan long sections of plant material. The implementation of larger image sensors will ameliorate some of these issues in the future and reduce the overall number of tiles necessary to capture an entire xylem network. Low resolution scans are usually sufficient to determine xylem conduit connectivity in (*i.e.* < 1 cm in diameter) provided the vessel diameters are large (*e.g.* 4.5 μm resolution used by Brodersen *et al.* 2012, 2013 to study grapevine vessels ~75–200 μm in diameter), while higher resolution imaging can reveal fine details like the location of intervessel wall pitting (*e.g.* Trtik *et al.* 2007; Van den Bulcke *et al.* 2008). Figure 3 shows the xylem network from the petiole of *Citrus sinensis* at 650 nm resolution. Because of the high resolution, both vessel-parenchyma and intervessel scalariform pitting are visible as well as the annulus delineating individual vessel elements (Fig. 3b, c). In a recent paper, Mayo *et al.* (2010) present images of the same wood sample visualized at three different resolutions and clearly show the advantages of each.

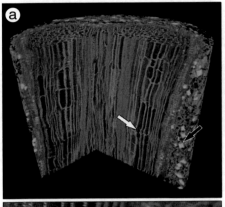
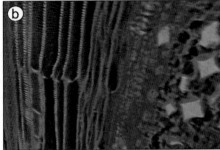
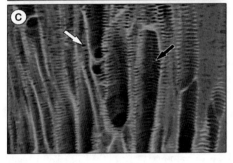

Figure 3. Three-dimensional volume rendering of a *Citrus sinensis* petiole visualized with µCT imaging (a) performed at the LBNL Advanced Light Source in Berkeley, CA, U.S.A. (Beamline 8.3.2). When visualized at 600 nm resolution, vessel wall details are visible including scalariform pitting and the annulus delineating vessel elements (white arrow). Calcium oxalate crystals are also visible (black arrow) embedded in the cortex. Panel (b) reveals a higher magnification region of (a). Scale varies with perspective, but the cylinder of tissue visualized in (a) is 1.7 mm in diameter, and the vessels in (b) are 20 µm in diameter. In (c), scalariform intervessel pitting is clearly visible (black arrow) as well as vessel-parenchyma pitting (white arrow).

EXAMPLES OF MCT IMAGING APPLICATIONS FOR WOOD ANATOMISTS

For wood anatomists, µCT has proven to be a highly useful tool that will aid in answering questions about the three-dimensional organization of the xylem and its relationship to the surrounding tissues. High resolution imaging systems now provide sufficient resolving power to clearly visualize the xylem in species bearing tracheids or small-diameter vessels and capturing the fine details of the conduit walls that are otherwise obscured with lower resolution scans (*e.g.* Fig. 3; Mayo *et al.* 2010). Of equal importance is the simplicity by which wood can be virtually dissected using µCT, and this technology will help wood anatomists to better understand the spatial arrangement of the xylem conduits and the supporting tissues. Recent research has shown that paratracheal parenchyma cells are of particular importance in the dynamic process of drought recovery (Tyree *et al.* 1999; Salleo *et al.* 2004; Brodersen *et al.* 2010; Brodersen & McElrone 2013), and µCT has proven to be an important tool in understanding this relationship. While we are endowed with a thorough knowledge of wood anatomy owing to the rich history of histologic wood preparations visualized with light microscopy and other techniques (Schweingruber 1990, 2011; Carlquist 2001; Tyree & Zimmermann 2002; Evert 2006), the three-dimensional relationships between tissue types in wood remain difficult to visualize with traditional techniques. µCT offers an additional, complementary tool to traditional methods.

For example, Robert *et al.* (2011) utilized µCT imaging to study the network structure of the xylem and phloem in *Avicennia* wood, revealing complex patterns of concentric, successive cambia. Such studies will help reveal the relationships between the structure and function of the xylem, particularly when framed in an ecological context. In another example, Page *et al.* (2011) used µCT to study the 3D organization of the xylem in co-occurring *Acacia* species from Australia. In this study the authors found that the degree of xylem connectivity was similar in all three species despite differences in branch water potential and other anatomical traits. Drought tolerance may be more tightly correlated with leaf shape than xylem anatomy within the leaf, providing yet another example of how selection can act at different levels of organization. Explorations into the 3D organization of the xylem show the future potential of µCT as a transformative tool that may further expand or redefine the wood anatomist's nomenclature as physiologically important structures become apparent.

One such example is the diversity of traits related to vessel grouping and how to determine from cross sections which vessels are connected through a shared wall (Carlquist 1984). Hass *et al.* (2010) used µCT to study the porosity of beech wood (presumably *Fagus sylvatica*) and identified groups ("clusters") of vessels that appeared to be connected within the block of wood. Connectivity was assumed when the vessels were in close enough proximity to each other or the two vessels appeared to merge. As the image resolution in the Hass *et al.* 2010 study was insufficient to visualize intervessel pitting, the problem arises of when to characterize vessels as being connected and whether vessels within a group are connected and over what axial length. One can imagine viewing a transverse section where vessels are grouped in pairs or triplets. Serial sectioning might reveal that those pairs exist for only a short axial distance and the connection is fleeting. Or, upon closer inspection, the group is within close proximity but not connected (Brodersen *et al.* 2011). With sufficient sampling vessel grouping characteristics can be obtained with confidence, and the total shared wall area and contact length between vessels can be approximated with serial light microscopy (Wheeler *et al.* 2005), a value that may be the much more important for water transport or drought resistance than the number of vessels in a group. The two characteristics are undoubtedly related, but as the 3D course of xylem conduits becomes better characterized in the future, wood anatomists will need to settle on a new set of characteristics with which to describe these three-dimensional relationships.

For example, how should we define the degree of contact between the xylem conduits and the ray parenchyma, particularly when we consider the organization in 3D? Here, the stem of a *Pinus taeda* seedling was scanned and visualized with µCT

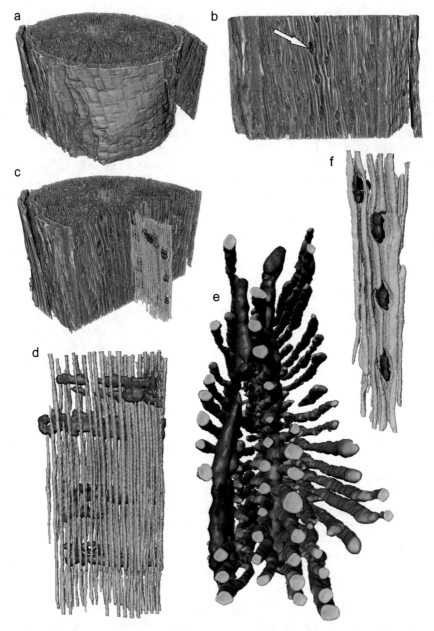

Figure 4. Three-dimensional µCT reconstruction of the stem of a *Pinus taeda* seedling scanned at 650 nm resolution. The volumetric rendering of the stem, approximately 1.2 mm in diameter was visualized as a whole (a), and was then virtually dissected to expose the a longitudinal plane (b) to expose some of the tracheids and the rays (arrow). In an alternate orientation, the rays and tracheids from a region of the sample were reconstructed from the 3D dataset (c). Once extracted, the 3D volumes can then be viewed from a wide range of perspectives (d, e, f) to reveal the spatial relationships between the two cell types. Scale varies with perspective.

imaging (Fig. 4). The 3D volume of the stem was visualized (Fig. 4a), and then dissected to expose a longitudinal plane through the wood revealing the tracheids and rays (Fig. 4b). Using computer software the tracheid network was reconstructed as well as the rays embedded in the xylem (Fig. 4c). Measuring the precise amount of ray-tracheid overlap with traditional light microscopy would be extremely difficult as the amount of overlap changes axially through the stem. Using μCT the images presented in Figure 4 were assembled in less than an hour using Octopus 8.3 (Institute for Nuclear Sciences, University of Ghent, Belgium) for the image reconstruction and Avizo 7.0 (VSG, Inc., Burlington, Massachusetts, USA) software for visualization (Brodersen *et al.* 2011), with an additional 45 minutes for scanning and mounting the sample. This method allows the ray-tracheid network to be visualized from a variety of different angles (Fig. 4d–f) and measuring the connectivity of the network is easily managed. Figure 4 represents a stem that is less than 1 mm in diameter and length, and reveals only a small fraction of the rays present in the sample. How much variability exists within the sample, throughout the whole stem, or between individual plants? By tiling several μCT scans together to capture a longer section of the stem and then broadly sampling from a population one could attempt to answer these questions.

At this early stage of our inquiries into 3D xylem organization using μCT it will be important to identify the relevant traditional characters (Wheeler *et al.* 1989) and whether they can continue to be informative when considering both 2D and 3D applications. Ideally, each characteristic wood anatomists use to characterize a wood specimen would be valid both in 2D and 3D. The vast majority of anatomical characters will remain valid, but the degree of connectivity between vessels, and other cell types, may necessitate a new set of descriptive terms to document these anatomical traits in 3D.

LIVE PLANT IMAGING

In addition to imaging dehydrated samples, intact plants can be visualized *in vivo* with μCT without compromising the tension on the xylem sap. This method has been used to study the spread of drought-induced embolism (Brodersen *et al.* 2013b) as well as the mechanism responsible for removing embolisms from the xylem network (Brodersen *et al.* 2010; Suuronen *et al.* 2013). While *in vivo* μCT imaging is still new, this method holds great promise for studying the functional status of intact xylem networks at resolutions that are significantly better than other non-destructive imaging tools (*e.g.* nuclear magnetic resonance (NMR) imaging; Holbrook *et al.* 2001). Studies specifically focused on visualizing the functional status of the xylem *in vivo* have broadened the utility of μCT and continue to emphasize the link between form and function. Currently, μCT and NMR are the two best methods available for monitoring the functional status of the xylem *in vivo*.

One of the biggest technological obstacles for future μCT studies in live plants is the dependence of image contrast on the X-ray attenuation differences between air and water. In hydrated tissue, the vessel lumen is easily visualized when filled with air, but the surrounding, water-filled tissue is largely obscured thereby making it difficult to distinguish the anatomy of the surrounding tissue (Brodersen *et al.* 2011). μCT images can be coupled with brightfield light microscopy or SEM images of the tissue following µCT scanning to get a comprehensive understanding of the spatial organization of the xylem. Finally, µCT has been used in combination with other methods for determining physiological parameters such as cavitation resistance, and allows for a visual confirmation of the spread of embolism that is otherwise measured using indirect methods (*e.g.* Choat *et al.* 2010; McElrone *et al.* 2012).

OTHER PLANT RELATED APPLICATIONS

µCT imaging has not been limited to the wood structure alone and has been employed in a wide range of applications related to plant biology. Dhondt *et al.* (2010) used µCT to study the morphological traits of *Arabidopsis thaliana* seedlings, including the trichomes on the leaf surface and detailed imaging of flower parts. Kaminuma *et al.* (2008) expanded on this line of research and used µCT to study the 3D anatomy of trichomes and their distribution on the surface of *A. thaliana* leaves, linking specific genes to the distribution of the trichomes on the leaf lamina, providing another method for phenotyping *A. thaliana* mutants. Low-resolution imaging can also be useful for studying the gross morphology of whole plant structures, such as the graft unions in stems (Milien *et al.* 2012), the anatomy of flower parts (Stuppy *et al.* 2003), or whole pieces of fruit (Verboven *et al.* 2008). Brodersen *et al.* (2012) used µCT to study the organization of the vascular bundles in two fern species, and by combining those images with ecophysiological tools were able to better understand the fundamental relationships between the structure and function of the xylem at a higher organizational level.

This technology has also been used successfully in the field of dendrochronology (*e.g.* Okochi *et al.* 2007; Grabner *et al.* 2009; Bill *et al.* 2012), with newer µCT instruments being capable of distinguishing between latewood and earlywood in tree rings. The ease of use and prevalence of traditional tree coring technology, however, appears to supersede the widespread use of µCT; however, automated image analysis and the three-dimensional capabilities could reveal additional characteristics that might be important to that field (Fonti *et al.* 2010). X-ray techniques have also found their way into the field of paleobotany (Boyce *et al.* 2003; DeVore *et al.* 2006; Smith *et al.* 2009; Scott *et al.* 2009), and due to the difficult nature of mineralized wood samples, µCT imaging may make previously intractable samples more amenable to research.

Recent advancements and combinations of different types of technology with µCT are promising and have shown the utility of this method for wood analysis from an industry perspective. For example, De Vetter *et al.* (2006) paired µCT with SEM X-ray spectroscopy to study cell wall penetration of chemical wood additives, Panthapulakkal and Sain (2013) studied changes in wood structure due to a chemical treatment, Taylor *et al.* (2013) used µCT to study wood shrinkage following dehydration, and Derome *et al.* (2011) found that µCT could be used to determine differences between earlywood and latewood shrinking and swelling in *Picea abies* wood samples. µCT technology is also being applied at the intersection of food science and wood anatomy, where Porter *et al.* (2011) used µCT to study the porosity of wood from different types of wine barrels. Forsberg *et al.* 2008 used µCT to study strain and wood deformation resulting in 3D displacement field diagrams. X-ray scattering and μCT were recently used to determine the differences in the microfibril orientations in the conduit walls of three different tree species (Svendström *et al*. 2012).

The technology also has the potential to significantly improve the mathematical models that have previously been based on xylem parameters picked from the literature to simulate a xylem network (*e.g*. Loepfe *et al*. 2007). By scanning entire vessel networks with μCT and then importing those three-dimensional data to populate the network models, water transport through the real xylem network can be simulated, including the response of the network to dysfunction resulting from drought-induced embolism, tylose formation, or the introduction of pathogens. Using this method, Lee *et al*. (2013) found that the inclusion or exclusion of xylem vessel relays (Brodersen *et al*. 2013a) from the network significantly impacts the redundancy of water transport, and under certain circumstances can generate scenarios where reverse (*i.e*. basipetal) flow is predicted. The identification of the spatial distribution of intervessel pitting should allow for more sophisticated modeling of water transport throughout xylem networks, and results from these simulations will be realistic and true to the original plant sample.

TECHNICAL CONSIDERATIONS AND POTENTIAL PITFALLS

Selecting an appropriate image resolution is of critical importance, as many of the current bottlenecks associated with μCT imaging are related to the large size of the image datasets. Resolution scales proportionally to the final dataset size, as the higher number of pixels in high-resolution images result in larger file sizes. Because the field of view decreases with increasing resolution, tiling is often required to visualize the whole sample. The datasets resulting from large tiling efforts can be substantial. For example, scanning a stem 8 mm in diameter over 5 mm imaged at 4.5 μm resolution following the methods of Brodersen *et al*. (2011) yields a stack of 1112 images, with a total size of 4–5 GB. The geometric scaling of data resulting from merging multiple tiles is sobering, particularly when considering the computer processing power, data storage, and graphics processing required to visualize the dataset as a whole. This tiling technique has been used to scan stem segments up to 4.5 cm in length (Brodersen *et al*. 2013a), but could be expanded to scan very long or wide segments. Theoretically, a tree trunk several centimeters in diameter could be scanned, with the resulting 3D image composed of hundreds of tiles. Each μCT instrument is different, and the dimensions of the sample stage and its range of motion will be the primary limitation to acquiring the data. Once the data is collected, the formidable task remains of merging the tiles into a continuous dataset and displaying it properly for analysis. At the time of writing, computers capable of handling such a dataset are exceedingly rare. However, computing power and data storage have decreased in cost during the past 10–15 years, but how long that industry can sustain this trend and follow Moore's law is uncertain (Schaller 1997; Mack 2011). The images presented here were created using a custom-built computer with 24 processors, 96 GB of RAM, and a dedicated high-end graphics processor. These computer systems are expensive, but costs are progressively decreasing and universities often have computers that are associated with the lab-based μCT systems.

A significant bottleneck in the μCT technique is image analysis. A variety of visualization software packages are available, both commercial and open-source, each with their own merits and the type of research will dictate the selection of an appropriate software package. This technology generates thousands of images, and developing a strategy to efficiently analyze the data is a significant challenge. Steppe *et al*. (2004) developed a custom computer program to automate μCT image analysis to measure vessel diameter, cross-sectional area of the xylem vessels, wood density, and other parameters. Similarly, Brodersen *et al*. (2011) expanded on that technique to automate the analysis of xylem anatomy and network connectivity. While the image preparation time was slightly more labor intensive using the automated software package (1 vs. 4 hours), the total amount of time required to analyze the same network information was significantly decreased (16 vs. 0.03 hours). The time was largely devoted to image processing which is dictated by computer processing power. Faster computers and new iterations of the software will accelerate this process, further reducing the amount of time necessary to analyze large datasets. Advances in image processing (*e.g.* Gil *et al*. 2009) will also improve the overall quality of the images prior to analysis with the aim of reducing image noise while preserving the inherent structures in the images.

CONCLUSIONS

As both the temporal and spatial resolution improve in future iterations of both lab-based and synchrotron μCT systems image quality will continue to improve. By far the most important key to future use of this technology is widespread access and collaborative research. Currently, lab-based systems are expensive and synchrotron-based systems have limited access. Focused, well-defined, hypothesis-driven studies that bring together wood anatomists and physiologists are likely to yield the most significant results with the limited resources available. Future studies should focus on obtaining images of larger samples, multiple samples from the same species or several species within a genus, and determining variability of functional wood characteristics across a broad range of species. Pairing anatomical analysis with physiological measurements will help to strengthen our understanding of the link between form and function. As noted above, readily available user-friendly software that aids in the automation of xylem network analysis will make the method much more attractive to users unfamiliar with the technique and help to standardize the measurements so that comparative studies are possible.

Finally, the highly visual nature of this technique has obvious implications for teaching plant anatomy. The 3D representations of xylem organization make the complexity of xylem networks less intimidating and it is easy to see the spatial relationships between the different cell types (*e.g.* Fig. 4). Instead of a two-dimensional image on a page, μCT brings the structures to life by giving them volume and form. Many of the 3D visualization software packages also allow components of the 3D models to be viewed as an interactive website with a typical student-level computer. Three-dimensional μCT images could also add a new facet to online wood databases such as InsideWood (insidewood.lib.ncsu.edu) and The Xylem Database (www.wsl.ch/dendro/xylemdb).

ACKNOWLEDGEMENTS

The author kindly thanks A. MacDowell and D. Parkinson for their assistance at the Lawrence Berkeley National Laboratory Advanced Light Source, Beamline 8.3.2 in Berkeley, CA, USA where the μCT imaging for this manuscript was performed. The Advanced Light Source is supported by the Director, Office of Science, Office of Basic Energy Services, of the US Department of Energy under contract no. DE-AC01-05CH11231. Dr. D. Johnson kindly provided the *Pinus taeda* seedling for μCT imaging. The author also thanks C. Manuck, M. Reed, and A. McElrone for their thoughtful edits to this manuscript.

REFERENCES

André J-P. 2005. Vascular organization of angiosperms: a new vision. Science Pub. Inc., Enfield, NH.

Beck C, Schmid R & Rothwell G. 1982. Stelar morphology and the primary vascular system of seed plants. Bot. Review 48: 691–815.

Bill J, Daly A, Johnsen Ø & Dalen K. 2012. DendroCT-Dendrochronology without damage. Dendrochronologia 30: 223–230.

Blonder B, Carlo F, Moore J, Rivers M & Enquist B. 2012. X-ray imaging of leaf venation networks. New Phytol. 196: 1274–1282.

Bosshard H & Kučera L. 1973. Die dreidimensionale Struktur-Analyse des Holzes – erste Mitteilung: Die Vernetzung des Gefäßsystems in *Fagus sylvatica* L. Holz Roh- Werkstoff 31: 437–445.

Boyce C, Cody G, Fogel M, Hazen R, Alexander C & Knoll AH. 2003. Chemical evidence for cell wall lignification and the evolution of tracheids in Early Devonian plants. Intern. J. Plant Sci. 164: 691–702.

Brodersen CR, Choat B, Chatelet D, Shackel KA, Matthews MA & McElrone AJ. 2013a. Xylem vessel relays contribute to radial connectivity in grapevine stems (*Vitis vinifera* and *V. arizonica*; Vitaceae). Amer. J. Bot. 100: 314–321.

Brodersen CR, Lee EF, Choat B, Jansen S, Phillips RJ, Shackel KA, McElrone AJ & Matthews MA. 2011. Automated analysis of three-dimensional xylem networks using high-resolution computed tomography. New Phytol. 191: 1168–1179.

Brodersen CR & McElrone AJ. 2013. Maintenance of xylem network transport capacity: a review of embolism repair in vascular plants. Frontiers in Plant Science 4.

Brodersen CR, McElrone AJ, Choat B, Lee EF, Shackel KA & Matthews MA. 2013b. In vivo visualizations of drought-induced embolism spread in *Vitis vinifera*. Plant Physiol. 161: 1820–1829.

Brodersen CR, McElrone AJ, Choat B, Matthews MA & Shackel KA. 2010. The dynamics of embolism repair in xylem: in vivo visualizations using high-resolution computed tomography. Plant Physiol. 154: 1088–1095.

Brodersen CR, Roark L & Pittermann J. 2012. The physiological implications of primary xylem organization in two ferns. Plant, Cell & Environm. 35: 1898–1911.

Bucur V. 2003. Nondestructive characterization and imaging of wood. Springer Verlag, Berlin.

Burggraaf P. 1972. Some observations on the course of the vessels in the wood of *Fraxinus excelsior* L. Acta Bot. Neerl. 21: 32–47.

Carlquist S. 1984. Vessel grouping in dicotyledon wood: significance and relationship to imperforate tracheary elements. Aliso 10: 505–525.

IAWA Journal 34 (4), 2013

Carlquist S. 2001. Comparative wood anatomy: systematic, ecological, and evolutionary aspects of dicotyledon wood. Ed. 2. Springer Verlag, Berlin.

Choat B, Drayton W, Brodersen C, Matthews M, Shackel A, Wada H & McElrone A. 2010. Measurement of vulnerability to water stress-induced cavitation in grapevine: a comparison of four techniques applied to a long-vesseled species. Plant, Cell & Environm. 33: 1502–1512.

De Vetter L, Cnudde V, Masschaele B, Jacobs P & Van Acker J. 2006. Detection and distribution analysis of organosilicon compounds in wood by means of SEM-EDX and micro-CT. Materials Characterization 56: 39–48.

Derome D, Griffa M, Koebel M & Carmeliet J. 2011. Hysteretic swelling of wood at cellular scale probed by phase-contrast X-ray tomography. J. Struct. Biol. 173: 180–190.

DeVore ML, Kenrick P, Pigg K & Ketcham R. 2006. Utility of high resolution X-ray computed tomography (HRXCT) for paleobotanical studies: an example using London Clay fruits and seeds. Amer. J. Bot. 93: 1848–1851.

Dhondt S, Vanhaeren H, Van Loo D, Cnudde V & Inzé D. 2010. Plant structure visualization by high-resolution X-ray computed tomography. Trends in Plant Science 15: 419–422.

Evert R. 2006. Esau's Plant anatomy: meristems, cells, and tissues of the plant body: their structure, function, and development. Wiley-Liss.

Fonti P, von Arx G, García-González I, Eilmann B, Sass-Klaassen U, Gärtner H & Eckstein D. 2010. Studying global change through investigation of the plastic responses of xylem anatomy in tree rings. New Phytol. 185: 42–53.

Forsberg F, Mooser R, Arnold M, Hack E & Wyss P. 2008. 3D Micro-scale deformations of wood in bending: synchrotron radiation μCT data analyzed with digital volume correlation. J. Struct. Biol. 164: 255–262.

Fujii T, Lee SJ, Kuroda N & Suzuki Y. 2001. Conductive function of intervessel pits through a growth ring boundary of *Machilus thunbergii*. IAWA J. 22: 1–14.

Gil D, Hernàndez-Sabaté A, Burnat M, Jansen S & Martínez-Villalta J. 2009. Structure-preserving smoothing of biomedical images. In: Computer Analysis of Images and Patterns: 427–434. Springer Verlag, Berlin.

Grabner M, Salaberger D & Okochi T. 2009. The need of high resolution μ-X-ray CT in dendrochronology and in wood identification. In: Zinterhof P, Uhl A, Loncaric S & Carini A (eds.), Image and Signal Processing and Analysis. ISPA 2009. Proceedings of 6th International Symposium: 349–352. IEEE.

Hass P, Wittel FK, McDonald SA, Marone F, Stampanoni M, Herrmann HJ & Niemz P. 2010. Pore space analysis of beech wood: the vessel network. Holzforschung 64: 639–644.

Holbrook N, Ahrens E, Burns M & Zwieniecki M. 2001. In vivo observation of cavitation and embolism repair using magnetic resonance imaging. Plant Physiol. 126: 27–31.

Huggett B & Tomlinson P. 2010. Aspects of vessel dimensions in the aerial roots of epiphytic Araceae. Intern. J. Plant Sci. 171: 362–369.

Kaminuma E, Yoshizumi T, Wada T, Matsui M & Toyoda T. 2008. Quantitative analysis of heterogeneous spatial distribution of *Arabidopsis* leaf trichomes using micro X-ray computed tomography. The Plant Journal 56: 470–482.

Kitin P, Fujii T, Abe H & Funada R. 2004. Anatomy of the vessel network within and between tree rings of *Fraxinus lanuginosa* (Oleaceae). Amer. J. Bot. 91: 779–788.

Kitin P, Sano Y & Funada R. 2003. Three-dimensional imaging and analysis of differentiating secondary xylem by confocal microscopy. IAWA J. 24: 211–222.

Lee E, Matthews M, McElrone A, Phillilps R, Shackel K & Brodersen C. 2013. Analysis of HRCT-derived xylem network reveals reverse flow in some vessels. J. Theor. Biol. (In press). http://dx.doi.org/10.1016/j.jtbi.2013.05.021

Loepfe L, Martinez-Vilalta J, Pinol J & Mencuccini M. 2007. The relevance of xylem network structure for plant hydraulic efficiency and safety. J. Theor. Biol. 247: 788–803.

Mack C. 2011. Fifty years of Moore's law. Semiconductor Manufacturing, IEEE Transactions on 24: 202–207.

Mayo S, Chen F & Evans R. 2010. Micron-scale 3D imaging of wood and plant microstructure using high-resolution X-ray phase-contrast microtomography. J. Struct. Biol. 171: 182–188.

McCulloh K, Sperry J & Alder F. 2003. Water transport in plants obeys Murray's law. Nature 421: 939–942.

McElrone A, Brodersen C, Alsina M, Drayton W, Matthews M, Shackel K, Wada H, Zufferey V & Choat B. 2012. Centrifuge technique consistently overestimates vulnerability to water stress-induced cavitation in grapevines as confirmed with high-resolution computed tomography. New Phytol. 196: 661–665.

Milien M, Renault-Spilmont A-S, Cookson S, Sarrazin A & Verdeil J-L. 2012. Visualization of the 3D structure of the graft union of grapevine using X-ray tomography. Scientia Horticulturae 144: 130–140.

Okochi T, Hoshino Y, Fujii H & Mitsutani T. 2007. Nondestructive tree-ring measurements for Japanese oak and Japanese beech using micro-focus X-ray computed tomography. Dendrochronologia 24: 155–164.

Oven P, Merela M, Mikac U & Sersa I. 2011. Application of 3D magnetic resonance microscopy to the anatomy of woody tissues. IAWA J. 32: 401–414.

Page G, Liu J & Grierson P. 2011. Three-dimensional xylem networks and phyllode properties of co-occurring *Acacia*. Plant, Cell & Environm. 34: 2149–2158.

Panthapulakkal S & Sain M. 2013. Investigation of structural changes of alkaline-extracted wood using X-ray microtomography and X-ray diffraction: A comparison of microwave versus conventional method of extraction. J. Wood Chem. & Technol. 33: 92–102.

Pittermann J, Brodersen C & Watkins Jr JE. 2013. The physiological resilience of fern sporophytes and gametophytes: advances in water relations offer new insights into an old lineage. Frontiers Plant Sci. 5: 285.

Porter G, Lewis A, Barnes M & Williams R. 2011. Evaluation of high power ultrasound porous cleaning efficacy in American oak wine barrels using X-ray tomography. Innovative Food Science & Emerging Technologies 12: 509–514.

Robert EM, Schmitz N, Boeren I, Driessens T, Herremans K, De Mey J, Van de Casteele E, Beeckman H & Koedam N. 2011. Successive cambia: a developmental oddity or an adaptive structure? PloS One 6: e16558.

Salleo S, Lo Gullo M, Trifilo P & Nardini A. 2004. New evidence for a role of vessel-associated cells and phloem in the rapid xylem refilling of cavitated stems of *Laurus nobilis* L. Plant, Cell & Environm. 27: 1065–1076.

Schaller R. 1997. Moore's law: past, present and future. Spectrum, IEEE 34: 52–59.

Schweingruber FH. 1990. Anatomy of European woods. Paul Haupt, Bern.

Schweingruber FH, Börner A & Schulze E-D. 2011. Atlas of stem anatomy in herbs, shrubs and trees. Springer Verlag, Berlin, Heidelberg.

Scott A, Galtier J, Gostling N, Smith S, Collinson M, Stampanoni M, Marone F, Donoghue P & Bengtson S. 2009. Scanning electron microscopy and synchrotron radiation X-ray tomographic microscopy of 330 million year old charcoalified seed fern fertile organs. Microscopy and Microanalysis 15: 166–173.

Smith S, Collinson M, Rudall P, Simpson D, Marone F & Stampanoni M. 2009. Virtual taphonomy using synchrotron tomographic microscopy reveals cryptic features and internal structure of modern and fossil plants. Proc. National Academy of Sciences 106: 12013–12018.

Steppe K, Cnudde V, Girard C, Lemeur R, Cnudde J & Jacobs P. 2004. Use of X-ray computed microtomography for non-invasive determination of wood anatomical characteristics. J. Struct. Biol. 148: 11–21.

Stuppy W, Maisano J, Colbert M, Rudall P & Rowe T. 2003. Three-dimensional analysis of plant structure using high-resolution X-ray computed tomography. Trends in Plant Science 8: 2–6.

Suuronen J-P, Peura M, Fagerstedt K & Serimaa R. 2013. Visualizing water-filled versus embolized status of xylem conduits by desktop X-ray microtomography. Plant Methods 9: 11.

Svedström K, Lucenius J, Van den Bulcke J, Van Loo D, Immerzeel P, Suuronen J-P, Brabant L, Van Acker J, Saranpää, P. & Fagerstedt K. 2012. Hierarchical structure of juvenile hybrid aspen xylem revealed using X-ray scattering and microtomography. Trees 26: 1793–1804.

Taylor A, Plank B, Standfest G & Petutschnigg A. 2013. Beech wood shrinkage observed at the micro-scale by a time series of X-ray computed tomographs (μXCT). Holzforschung 67: 201–205.

Trtik P, Dual J, Keunecke D, Mannes D, Niemz P, Stahli P, Kaestner A, Groso A & Stampanoni M. 2007. 3D imaging of microstructure of spruce wood. J. Struct. Biol. 159: 46–55.

Tyree M, Salleo S, Nardini A, Gullo M & Mosca R. 1999. Refilling of embolized vessels in young stems of laurel. Do we need a new paradigm? Plant Physiol. 120: 11–22.

Tyree MT & Zimmermann MH. 2002. Xylem structure and the ascent of sap. Springer, Berlin, New York.

Van den Bulcke J, Masschaele B, Dierick M, Van Acker J, Stevens M & Hoorebeke L. 2008. Three-dimensional imaging and analysis of infested coated wood with X-ray submicron CT. International Biodeterioration & Biodegradation 61: 278–286.

Verboven P, Kerckhofs G, Mebatsion HK, Ho QT, Temst K, Wevers M, Cloetens P & Nicolaï BM. 2008. Three-dimensional gas exchange pathways in pome fruit characterized by synchrotron X-ray computed tomography. Plant Physiol. 147: 518–527.

Wheeler EA, Baas P & Gasson P. 1989. IAWA list of microscopic features for hardwood identification. IAWA Bull. n.s. 10: 219–332.

Wheeler J, Sperry J, Hacke U & Hoang N. 2005. Inter-vessel pitting and cavitation in woody Rosaceae and other vesselled plants: a basis for a safety versus efficiency trade-off in xylem transport. Plant, Cell & Environm. 28: 800–812.

Wu H, Jaeger M, Wang M, Li B & Zhang B. 2011. Three-dimensional distribution of vessels, passage cells and lateral roots along the root axis of winter wheat (*Triticum aestivum*). Ann. Bot. 107: 843–853.

Zimmerman M. 1971. Dicotyledonous wood structure made apparent by sequential sections. In: Wolf G (ed.), Encyclopaedia cinematographica. Institut für den Wissenschaftlichen Film. Göttingen, Germany.

Zimmerman M & Brown C. 1971. Trees: structure and function. Springer-Verlag, Berlin, New York.

Zimmermann M & Tomlinson P. 1966. Analysis of complex vascular systems in plants: optical shuttle method. Science 152: 72–73.

Zimmermann M & Tomlinson P. 1968. Vascular construction and development in the aerial stem of *Prionium* (Juncaceae). Amer. J. Bot. 55: 1100–1109.

Accepted: 8 August 2013

BRILL

IAWA Journal 34 (4), 2013: 425–432

ROXAS – AN EFFICIENT AND ACCURATE TOOL TO DETECT VESSELS IN DIFFUSE-POROUS SPECIES

Lena Wegner[1,*], **Georg von Arx**[2], **Ute Sass-Klaassen**[1] and **Britta Eilmann**[1]

[1]Forest Ecology and Forest Management Group, Centre for Ecosystem Studies,
Wageningen University, PO Box 47, 6700 AA Wageningen, The Netherlands
[2]Swiss Federal Institute for Forest, Snow and Landscape Research WSL, Landscape Dynamics
Unit, Zürcherstr. 111, CH-8903 Birmensdorf, Switzerland
*Corresponding author; E-mail: lena.wegner@wur.nl

ABSTRACT

Wood-anatomical parameters form a valuable archive to study past limitations on tree growth and act as a link between dendrochronology and ecophysiology. Yet, analysing these parameters is a time-consuming procedure and only few long chronologies exist. To increase measurement efficiency of wood-anatomical parameters, novel tools like the automated image-analysis system ROXAS were developed. So far, ROXAS has only been applied to measure large earlywood vessels in ring-porous species.

In this study, we evaluate if ROXAS is also suitable for efficient and accurate detection and measurement of vessels in diffuse-porous European beech. To do so, we compared the outcome of ROXAS with that of the established measurement programme Image-Pro Plus in terms of efficiency and accuracy.

The two methods differed substantially in efficiency with automatic measurements using ROXAS being 19 times faster than with Image-Pro Plus. Although the procedures led to similar patterns in annual variation of mean vessel area and vessel density, the absolute values differed. Image-Pro Plus measured systematically lower mean vessel areas and higher vessel densities than ROXAS. This was attributed to the species-specific technical settings in ROXAS, leading to more realistic results than those obtained using the default settings in Image-Pro Plus. A shortcoming of ROXAS was, however, that small vessels ($<100 \ \mu m^2$) could not be detected with sufficient accuracy. Yet, based on thin sections of European beech, it is generally difficult to distinguish such small vessels from parenchyma cells. Moreover, these small vessels do not contribute substantially to conductive efficiency. Therefore, we do not foresee any problems for most studies if the lower vessel area threshold to be measured is set to 100 μm^2. Overall, ROXAS proved to be useful for measuring vessel parameters in diffuse-porous tree species, allowing accurate and efficient analyses of large numbers of samples.

Keywords: Wood anatomy, automatic vessel detection, novel techniques, robust chronologies.

DOI 10.1163/22941932-00000034

INTRODUCTION

Trees form a valuable archive of environmental conditions that determined tree growth, since variation in climatic conditions leave permanent imprints on the wood structure (Schweingruber 1996; Fonti *et al.* 2010). Especially ring width is widely used as a powerful and reliable proxy to study climate impacts on tree growth (*e.g.* Briffa *et al.* 2004). However, tree rings show an integrated response to climate conditions in the previous and the current year (Fritts 2001), making the interpretation of climate-growth relationships challenging (Kagawa *et al.* 2006). In contrast, wood-anatomical parameters can potentially reflect climate conditions at the time of their formation which allows studying climate-growth relationships with higher time resolution (García-González & Eckstein 2003; Fonti & García-González 2004). Moreover, wood-anatomical parameters, such as vessel size, are strongly linked to the trees' metabolism and reveal consequences of climate limitations on tree growth. Thus, wood-anatomical parameters can act as a link between retrospective dendrochronology providing long-time series and detailed but short-term analysis of ecophysiological parameters (Fonti *et al.* 2010).

Despite the many potential advantages of wood-anatomical parameters, long wood-anatomical chronologies are not as widely established as tree-ring chronologies, which is mainly due to the time-consuming analysis. Thus, few long wood-anatomical chronologies exist, often focussing on big earlywood vessels only (*e.g.* Eilmann *et al.* 2009). Chen and Evans (2005, 2010) proposed a method for the automated measurement of vessel properties in some diffuse-porous tree species, using digital microscopy images obtained by transmitted red light. To increase measurement efficiency for wood-anatomical parameters while depending on the established image analysis programme Image-Pro Plus (Media Cybernetics, Silver Spring, Maryland, USA; in the following IPP), the semi-automatic image analysis tool ROXAS was developed (von Arx & Dietz 2005; von Arx *et al.* 2013). Based on species-specific settings, ROXAS automatically recognises and measures cells and annual rings, produces a suit of derived parameters, and saves all data into an MS Excel file (see von Arx 2013 for more details). ROXAS already proved to be suitable for measuring large earlywood vessels in oak with sufficient efficiency and accuracy (Fonti *et al.* 2009). However, it remained unclear whether ROXAS can be applied for vessel measurements in diffuse-porous tree species.

In this study, we tested if ROXAS is able to measure vessel parameters in diffuse-porous European beech (*Fagus sylvatica* L.) with sufficient accuracy and efficiency. We measured vessel size and vessel density in nine consecutive tree rings with three different methods: (i) using ROXAS without and (ii) with manual refinement of the automatic measurement output, and (iii) using IPP with manual refinement after automatic measurement. The output of these methods was compared in terms of measurement accuracy and time efficiency.

MATERIAL AND METHODS

Sampling and cross-dating of tree rings

We analysed nine consecutive tree rings from a dominant 13-year-old European beech tree growing in a provenance trial near Wageningen, the Netherlands (latitude

51° 97' N, longitude 5° 70' E). The 5 mm diameter increment core was taken at breast height in summer 2011 and covered the entire diameter of the tree. Nine additional trees of the same provenance were sampled to produce a ring-width reference chronology. All ten increment cores were prepared with a core microtome (WSL Birmensdorf, Switzerland).

Tree-ring widths were measured using Lintab (Rinntech Heidelberg, Germany) and visually cross-dated to assign each ring to the corresponding calendar year (TSAP; Rinntech Heidelberg, Germany).

Preparation of thin sections

To analyse vessel size and density, we prepared thin sections with a thickness of 5 μm by using a sliding microtome (WSL Birmensdorf, Switzerland). The thin sections were stained with a mixture of astrablue and safranin (150 mg astrablue, 40 mg safranin and 2 ml acetic acid in 100 ml distilled water), dehydrated with a gradient of ethanol (50, 90 and 100%) and xylol, and finally embedded permanently in Canada balsam (Table 1).

Image acquisition

Micro-images were taken (with the software Leica Application Suite v3.8) using a camera (Leica DFC320) installed on an optical microscope (Leica DM2500) and connected to a computer. To obtain images covering the whole increment core with a reasonable resolution, several overlapping photographs (taken with a 2.5× objective, image resolution: 300 dpi) were merged using the photo stitching software PTGui (v9.1.2) (Table 1).

Table 1. Overview of the procedures and their requirements for measuring vessels in one core (9 annual rings).

Procedures	ROXAS$_{auto}$	ROXAS$_{adj}$	IPP
Preparation			
Processes	Thin-sectioning, staining, embedding, oven treatment		
Time	14 h including 12 h oven treatment		
Image acquisition			
Number of images	2–6 per ring		
Objective	2.5×		
Resolution	300 dpi		
Editing	Stitching with PTGui		
Time	50 min		
Measurement			
Step unit	Whole core	Whole core	Ring by ring
Mode	Automatic	With manual correction	With manual correction
Time[1]	45 min	9 h	14 h

[1] Measurement times serve as an indication only.

Comparison of measurement procedures

Vessel size and vessel density were analysed using three different procedures: (i) ROXAS automatic analysis (von Arx & Dietz 2005; in the following ROXAS$_{auto}$), (ii) ROXAS automatic analysis followed by in-built manual refinement functionality (in the following ROXAS$_{adj}$), and (iii) analysis with the software IPP as an established procedure (Table 1). The time required for the measurements with the three different methods was recorded.

Both ROXAS procedures initially enhanced image quality in several steps (for details see von Arx & Dietz 2005). Vessels were then automatically detected using species-specific settings for European beech and the used staining and image-acquisition method. These settings included among others colour-, size- and shape-sensitive rules to discriminate vessels from lignified parenchyma cells and image noise. Before vessel extraction, merged candidate vessels were split. The tree-ring borders were drawn manually since an automatic separation of tree rings based on the species-specific settings was unreliable owing to the small gradient in vessel size and vessel density. After analysis, ROXAS automatically saved all data into an MS Excel file. Although the ROXAS output provided a large number of derived parameters, ranging from the number of vessels to potential hydraulic conductivity, we focussed on mean vessel area (MVA) and vessel density (VD).

For ROXAS$_{adj}$, the automatic detection and measurement of vessels was followed by manual correction of misidentified or unrecognised vessels using an in-built editing toolbox. Since deleting misidentified vessels was less time consuming than manually outlining unrecognised vessels, the colour and shape settings in ROXAS$_{adj}$ for extracting vessels were deliberately chosen to be less restrictive than in ROXAS$_{auto}$.

For the analysis with IPP, we used the default settings. Concerning minimum and maximum vessel size, the same settings were used as for the two ROXAS procedures. Analyses were done ring by ring, using the functions "automatic [recognition of] bright objects" and "auto split". Measurement errors were corrected manually by splitting clustered vessels, deleting misrecognised vessels and drawing unrecognised vessels.

The Gleichläufigkeitskoeffizient (GLK; Eckstein & Bauch 1969) – which reflects the percentage of oscillations in the same direction in two time series within a certain period – between MVA and VD measurements of the three tested methods was calculated to assess the accuracy of the two ROXAS measurements. Furthermore, the MVA values were used to calculate Pearson's correlation coefficient (r) between the IPP measurements and the two ROXAS measurements. For this purpose, we gradually increased the minimum vessel area to be included from 40 to 6000 μm^2 and calculated the MVA of every tree ring for the three different methods.

RESULTS AND DISCUSSION

The time required for analysis of the sample was nearly 19 times longer when using IPP in comparison to ROXAS$_{auto}$ and still 1.5 times longer than with ROXAS$_{adj}$ (Table 1). However, regarding the patterns in MVA and VD the results were quite similar for all three procedures (Fig. 1) with a GLK value of 0.75 between IPP and both ROXAS

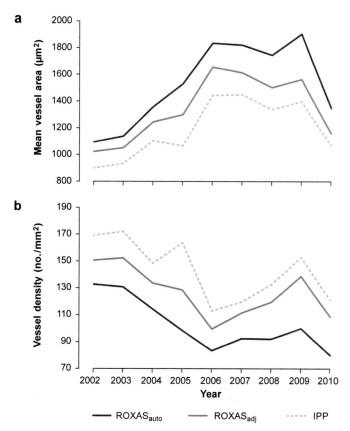

Figure 1. **a**: Mean vessel area and **b**: vessel density measured with ROXAS$_{auto}$, ROXAS$_{adj}$ and IPP.

measurements for MVA and GLK values of 0.88 and 0.63 between IPP and ROXAS$_{adj}$ and IPP and ROXAS$_{auto}$, respectively, for VD. Still, the absolute values of MVA and VD differed systematically. IPP yielded the lowest absolute MVA values of all procedures (Fig. 1a). MVA values measured with ROXAS$_{adj}$ were on average 13 % larger than with IPP, while MVA values measured with ROXAS$_{auto}$ were on average 28 % larger than with IPP. The low absolute MVA values in IPP are mainly attributed to an exclusion of shadows from the cell wall in the measurements as a result of a more restrictive threshold during vessel identification. This led to a systematic underestimation of vessel size (Fig. 2a & b). Thus, owing to the species-specific settings, the two ROXAS methods likely provide more realistic results for vessel size than the manual measurement with IPP.

The highest vessel density was detected when using IPP (Fig. 1b). VD values of ROXAS$_{auto}$ and ROXAS$_{adj}$ were on average 28 % and 11 % smaller than in IPP, owing to a better recognition of small objects in IPP. To determine the lower size threshold for an accurate detection of vessels with ROXAS, we correlated the output of IPP and

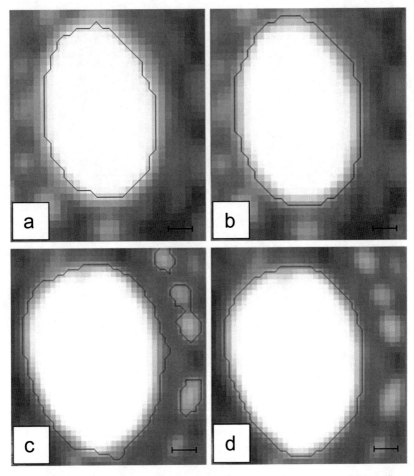

Figure 2. Different threshold values for the identification of vessels in (a) IPP and (b) ROXAS$_{auto}$. Parenchyma cells identified as vessels in (c) IPP were correctly excluded with (d) ROXAS$_{auto}$ due to the shape criteria. — Scale bars = 10 µm.

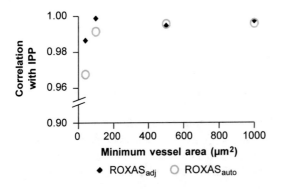

Figure 3. Correlation between mean vessel area values measured with IPP and ROXAS$_{auto}$ and IPP and ROXAS$_{adj}$ depending on the minimum vessel area.

the two ROXAS measurements and found that only very small vessels (area <100 μm^2) were seemingly detected more accurately with IPP. For a minimum vessel area of 100 μm^2, correlation coefficients stabilised around 0.999 for IPP and ROXAS$_{adj}$ and 0.991 for IPP and ROXAS$_{auto}$ (Fig. 3). However, for European beech, it is almost impossible to distinguish vessels smaller than 100 μm^2 from parenchyma cells of similar size and shape in thin sections. Thus, the more accurate measurement with IPP below 100 μm^2 might also be pseudo-exactness. For example, in 2005, the MVA curve based on IPP measurements showed a strong drop compared to the two ROXAS measurements (Fig. 1a). This was due to the detection of a large number of very small objects which are most likely parenchyma cells. However, owing to the similarity in cell wall structure between small vessels and parenchyma cells, the detection of these objects could not be corrected during manual editing (Fig. 2c & d). This resulted in an overestimation of VD for IPP (Fig. 1b). Owing to these discrimination problems, we advise to increase the lower vessel area measurement threshold to 100 μm^2 in order to avoid measurement errors by parenchyma cells. Also from an ecophysiological point of view these small vessels might be negligible, since they do not contribute substantially to conductive efficiency (Sass & Eckstein 1995; Fonti *et al.* 2010). Moreover, based on the output of IPP in our sample, objects identified as vessels smaller than 100 μm^2 represented on average only 8.5 % of the total number of vessels (SD = 2.9) and 0.47 % of the total vessel area (SD = 0.19).

Comparing the two ROXAS measurements, ROXAS$_{adj}$ produced a more accurate output than ROXAS$_{auto}$, as misrecognised vessels were corrected manually after automatic analysis. However, analyses with ROXAS$_{auto}$ were substantially faster than analyses with ROXAS$_{adj}$ (Table 1). Thus, measurement errors in ROXAS$_{auto}$ are partly compensated for by the large number of vessels that can be efficiently measured with this method. Hence, for comparing tree rings measured with the same measurement method, ROXAS$_{auto}$ gives sufficient accuracy and the highest efficiency compared to IPP and ROXAS$_{adj}$.

CONCLUSION

ROXAS is a suitable tool for the analysis of wood-anatomical parameters in diffuse-porous beech, measuring vessel size as well as vessel density with sufficient accuracy. Using species-specific settings, ROXAS even provided a more realistic output than IPP with its default settings. In addition, ROXAS$_{auto}$ was much more efficient than the manual analysis methods. In summary, ROXAS can be used for the development of long and robust chronologies of wood-anatomical parameters.

ACKNOWLEDGEMENTS

We would like to thank Sven de Vries from Alterra and the EU COST project E52 'Evaluation of Beech Genetic Resources for Sustainable Forestry' for providing us with information about and access to the field trial and the beech material. We are grateful to Leo Goudzwaard for support in the field and to Ellen Wilderink for support in the laboratory. This study was conducted in the framework of the COST Action FP1106, STReESS and was funded by the Marie Curie grant for intra-European fellowships for career development within the framework of the project INPUT-drought (FP7-274085).

REFERENCES

Briffa KR, Osborn TJ & Schweingruber FH. 2004. Large-scale temperature inferences from tree rings: a review. Global and Planetary Change 40: 11–26.

Chen F & Evans R. 2005. A robust approach for vessel identification and quantification in eucalypt pulpwoods. Appita J. 58: 442–447.

Chen F & Evans R. 2010. Automated measurement of vessel properties in birch and poplar wood. Holzforschung 64: 369–374.

Eckstein D & Bauch J. 1969. Beitrag zur Rationalisierung eines dendrochronologischen Verfahrens und zur Analyse seiner Aussagesicherheit. Forstwiss. Centralblatt 88: 230–250.

Eilmann B, Zweifel R, Buchmann N, Fonti P & Rigling A. 2009. Drought-induced adaptation of the xylem in Scots pine and pubescent oak. Tree Physiol. 29: 1011–1020.

Fonti P, Eilmann B, García-González I & von Arx G. 2009. Expeditious building of ring-porous earlywood vessel chronologies without losing signal information. Trees 23: 665–671.

Fonti P & García-González I. 2004. Suitability of chestnut earlywood vessel chronologies for ecological studies. New Phytol. 163: 77–86.

Fonti P, von Arx G, García-González I, Eilmann B, Sass-Klaassen U, Gärtner H & Eckstein D. 2010. Studying global change through investigation of the plastic responses of xylem anatomy in tree rings. New Phytol. 185: 42–53.

Fritts HC. 2001. Tree rings and climate. Blackburn Press, New Jersey.

García-González I & Eckstein D. 2003. Climatic signal of earlywood vessels of oak on a maritime site. Tree Physiol. 23: 497–504.

Kagawa A, Sugimoto A & Maximov TC. 2006. Seasonal course of translocation, storage and remobilization of C-13 pulse-labeled photoassimilate in naturally growing *Larix gmelinii* saplings. New Phytol. 171: 793–804.

Sass U & Eckstein D. 1995. The variability of vessel size in beech (*Fagus sylvatica*) and its eco-physiological interpretation. Trees – Structure and Function 9: 247–252.

Schweingruber FH. 1996. Tree rings and environment – dendroecology. Paul Haupt, Bern.

von Arx G. 2013. http://www.wsl.ch/roxas.

von Arx G & Dietz H. 2005. Automated image analysis of annual rings in the roots of perennial forbs. Intern. J. Plant Sci. 166: 723–732.

von Arx G, Kueffer C & Fonti P. 2013. Quantifying plasticity in vessel grouping – added value from the image analysis tool ROXAS. IAWA J. 34: 433–445.

Accepted: 13 August 2013

QUANTIFYING PLASTICITY IN VESSEL GROUPING – ADDED VALUE FROM THE IMAGE ANALYSIS TOOL ROXAS

Georg von Arx[1,*], Christoph Kueffer[2] and Patrick Fonti[1]

[1]Swiss Federal Institute for Forest, Snow and Landscape Research WSL, Birmensdorf, Switzerland
[2]Swiss Federal Institute of Technology ETH, Zurich, Switzerland
*Corresponding author; e-mail: georg.vonarx@wsl.ch

ABSTRACT

The functional role of the connectivity of the xylem network, especially the arrangement of solitary and grouped vessels in a cross section, has often been discussed in the literature. Vessel grouping may improve hydraulic integration and increase resilience to cavitation through redundancy of hydraulic pathways. Alternatively, a high degree of hydraulic integration may facilitate the spread of cavitations among neighboring vessels. Here we show how automated image analysis tools such as ROXAS (see www.wsl.ch/roxas) may greatly enhance the capacity for studying vessel grouping while avoiding some methodological limitations of previous approaches. We tested the new analysis techniques by comparing the xylem network of two populations of the herbaceous species *Verbascum thapsus* collected at a dry and moist site on Big Island (Hawaii, USA). ROXAS accurately, objectively and reproducibly detected grouped and solitary vessels in high-resolution images of entire root cross sections, and calculated different and partly novel vessel grouping parameters, *e.g.* the percentage of grouped (*vs.* solitary) vessels among different vessel size classes. Individuals at the dry site showed a higher degree of vessel grouping, less solitary vessels, greater maximum vessel sizes and an increase of the percentage of grouped vessels with increasing vessel size. The potential, but also some limitations of automated image analysis and the proposed novel parameters are discussed.

Keywords: Hydraulic pathways, automated image analysis, grouped vessels, hydraulic integration, solitary vessels, spreading of embolism, spatial vessel arrangement, vessel grouping index.

INTRODUCTION

The connectivity among vessels, also referred to as vessel grouping in a cross-sectional view (Carlquist 1984; Loepfe *et al.* 2007; Carlquist 2009), is increasingly attracting research interest as a potentially important hydraulic trait in angiosperms, particularly in a context of changing climate. This is because of the outstanding importance of vessels for water-transport. It is widely known, for example, that efficiency potentially increases with the fourth power of vessel radius according to the Hagen-Poiseuille law (Tyree & Ewers 1991), but also that wider vessels are arguably more susceptible to hydraulic

© International Association of Wood Anatomists, 2013
Published by Koninklijke Brill NV, Leiden

DOI 10.1163/22941932-00000035

failure by drought-induced cavitation, *i.e.* the rapid phase change of liquid water to vapor ("safety *vs.* efficiency trade-off"; *e.g.* Hacke & Sperry 2001; Hacke *et al.* 2009; Cai & Tyree 2010). The importance of the topology of vessel networks for plant water transport and plant-water relationships in general lies in the water-permeable pit pores formed in the cell walls between neighboring vessels. Previous studies observed an evolutionary trend towards an increase of vessel grouping with increasing water limitation (Carlquist 1966; Lopez *et al.* 2005; Lens *et al.* 2011; Carlquist 2012), although some studies also found inconsistent patterns or no relationship (Baas & Carlquist 1985; Baas & Schweingruber 1987). Similarly, a consistent increase of vessel grouping in response to water limitation and wounding was observed within species (Robert *et al.* 2009) and individuals (Arbellay *et al.* 2012), respectively.

However, the functional role of vessel grouping remains controversial. Several scientists suggested that a high degree of vessel grouping may provide alternative pathways when water transport through a vessel is blocked by drought-induced embolism but bypassed through one or more still functional vessels from the same vessel group. (Baas *et al.* 1983; Zimmermann 1983; Carlquist 1984; Tyree *et al.* 1994). In this case vessel grouping improves the hydraulic redundancy and reduces the potential loss of water transport capacity associated with cavitation. Another benefit of vessel grouping is related to the observed increasing permeability of intervessel pit membranes upon changes in the ionic concentration of the xylem sap which can occur during water limitation ("ionic effect", Jansen *et al.* 2011; Nardini *et al.* 2012). Finally, cavitations may be more easily removed in clustered than in solitary vessels, although this has so far only been postulated theoretically (Hölttä *et al.* 2006). In contrast, an increase of vessel grouping can also bring disadvantages. For example, vessels in a dense network have more contact surface, which enhance the risk of cavitation spreading from one vessel to the next by the aspiration of air through the pit pores (Sperry & Tyree 1988; Brodersen 2013; Brodersen *et al.* 2013) known as air-seeding hypothesis (Alder *et al.* 1997; Wheeler *et al.* 2005; Loepfe *et al.* 2007). Furthermore, vasicentric tracheids, where present, potentially offer a subsidiary conductive tissue that would lower the value of hydraulic integration by vessel grouping (Carlquist 1984, 2001), but likely not invalidate it (Sano *et al.* 2011). The empirical evidence from the relatively few quantitative studies investigating vessel grouping is insufficient to clarify the controversy about its functional role under water limitation.

Vessel grouping has been often quantified visually from the xylem cross section (see Mencuccini *et al.* 2010; Martínez-Vilalta *et al.* 2012 for some exceptions). However, this approach bears several potential limitations. First and most importantly, the coverage is usually restricted to a rather small sub-area of the entire sample and is not representative of the whole variability (see Arbellay *et al.* 2012 for an exception). Secondly, the results are hardly reproducible because there is usually some interpretation involved as to whether two vessels are truly connected through pit pores or not. Thirdly, the possibilities to quantify different aspects of the connectivity of the xylem network beyond grouped *vs.* solitary vessels and group size are limited. These methodological shortcomings can be significantly reduced through the use of image analysis tools for automatic detection and quantification of conduits. Advancements in digital imaging and improved computer performance allow nowadays to perform analysis of large samples while using objective criteria for the definition of vessel grouping. In this article, we present a new approach for quantifying several and partly novel parameters of vessel grouping in entire cross sections and tree cores using the image analysis tool ROXAS (von Arx & Dietz 2005; Fonti *et al.* 2009; Wegner *et al.* 2013). We test the potential of ROXAS for whole-sample grouping analysis with data of the forb *Verbascum thapsus* collected at two sites with contrasting climate, and suggest some directions for future studies into vessel grouping.

MATERIALS AND METHODS

Test dataset

To test the potential of whole-sample grouping analysis, and to investigate the plasticity of several aspects of vessel grouping within a species, we collected roots of 14 individuals of *Verbascum thapsus* L. (Scrophulariaceae) at two contrasting sites on Big Island, Hawaii, USA, between March 28th and April 24th 2008. The two sites represent the dry and wet margin of the distribution of *V. thapsus* in the selected study area (C. Kueffer, unpublished data). The first site ('dry' site), was at 600 m asl on a north-westerly, leeward slope (mean annual temperature: 20.5 °C, total annual precipitation: 790 mm, n = 6). The second site ('moist' site) was at 1700 m asl on an easterly, windward slope (14.5 °C, 2380 mm, n = 8). Plants at both sites grew on lava gravel characterized by poorly developed soil and sparse vegetation cover. The two sites were 60 km apart. The study species *V. thapsus* is a stout, erect herbaceous species that generally produces a rosette in the first and a flowering stalk in the second growing season, after which it dies. *Verbascum thapsus* is considered a problematic invasive species (Kueffer *et al.* 2010), and it is known to show very high plasticity in aboveground growth patterns in response to environmental variability (Parker *et al.* 2003; Seipel *et al.* 2013). All selected individuals were flowering when roots were collected. Plants from the dry site were 1.30 ± 0.15 m tall and plants from moist site 0.64 ± 0.08 m.

After collection, the proximal part of the main root of each individual was conserved in a 50 % ethanol solution. Complete 30–50 μm thick cross sections near the proximal end (root collar) were produced using a sledge microtome and stained with phloro-glucinol/HCl causing reddish coloration of lignified structures (here the cell walls of secondary xylem vessels and lignified parenchyma cells). Stained cross sections were then photographed through the oculars of a transmitting microscope (Olympus BX51, 20× magnification) using a standard digital camera (Nikon Coolpix 990). Multiple overlapping images were taken from each sample and stitched together using PTGui (New House Internet Services BV, Rotterdam, NL) to obtain high-resolution images (0.477 pixels/μm) of the entire cross sections (Fig. 1a, b; von Arx *et al.* 2012)). Digital images of cross sections were then analyzed using ROXAS, and automatic grouping results were compared to manual inspection of the digital images in three randomly chosen individuals. Manual inspection was performed by systematically counting the vessel groups of different sizes in the digital images. Only vessels from the root sections representing the flowering year in 2008 according to herb-chronological analysis were considered for analysis (von Arx & Dietz 2006), and from these only vessels having a

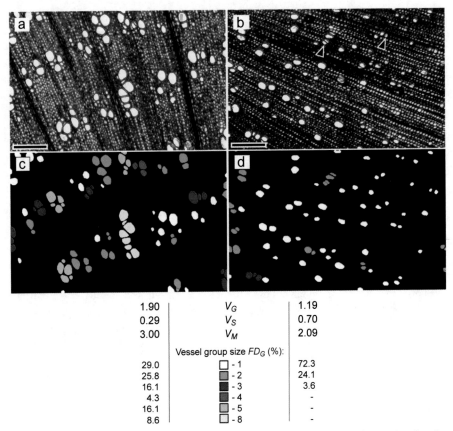

1.90	V_G	1.19
0.29	V_S	0.70
3.00	V_M	2.09

Vessel group size FD_G (%):

29.0	☐ - 1	72.3
25.8	▨ - 2	24.1
16.1	■ - 3	3.6
4.3	■ - 4	-
16.1	☐ - 5	-
8.6	☐ - 8	-

Figure 1. Root cross sections (cut-out images) of *Verbascum thapsus* from a dry (a, c) and moist site (b, d) showing how the image analysis tool ROXAS automatically quantifies vessel grouping patterns. – a & b: Original images with outlines of considered vessels in red. – c & d: Binary images of considered vessels with vessels belonging to different vessel group sizes depicted in different colors and values of several vessel grouping parameters given below. The left arrow in b indicates a paired vessel that might actually be overlapping ends of two vessels. The right arrow in b shows a vessel that was excluded because of the lower cut-off vessel size. See Table 1 for explanations of acronyms of vessel grouping parameters. — Scale bars in a and b = 200 μm.

cross-sectional lumen area ('vessel size') ≥ 250 μm², because below this size vessels could not be unambiguously distinguished from parenchyma cells.

Parameters to quantify vessel grouping

From a functional perspective, there is no unique best parameter to quantify vessel grouping, because several properties become relevant depending on the ecophysiological processes considered. We therefore considered five different parameters to quantify vessel grouping (Table 1). The first parameter was the vessel grouping index V_G proposed by Carlquist (2001), *i.e.* the mean number of vessels per vessel group (counting a solitary vessel as 1, a pair of vessels in contact as 2, etc.), which also corresponds

Table 1. Parameters of vessel grouping calculated with the image analysis tool ROXAS.

Acronym	Definition
V_G	Vessel grouping index; mean number of vessels per group (counting a solitary vessel as 1, a pair of connected vessels as 2, etc.; Carlquist 2001)
V_S	Vessel solitary fraction; ratio of solitary vessels to all vessels
V_M	Mean group size of non-solitary vessels
FD_G	Frequency distribution of different group sizes (absolute or relative)
RG_{VA}	Percentage of grouped (non-solitary) vessels per vessel size class

to the most commonly used parameter in the literature. While V_G provides a good general estimate of vessel grouping, it misses information about variation. Three other parameters were thus defined: the fraction of solitary vessels (V_S), the mean group size of non-solitary vessels (V_M), and the frequency distribution of vessel groups (FD_G). The rationale behind using V_G, V_S, and V_M is that two individuals may have the same value of V_G, but one of them may have smaller vessel groups (V_M) and thus fewer solitary vessels (V_S), than the other (Fig. 2). Compared to V_G and V_M, FD_G better represents the range of vessel grouping and the relative importance of different group sizes for overall connectivity. Finally, the functional meaning of vessel grouping may not be the same for wide vessels (expected to cavitate first) as for narrow vessels (expected to provide an hydraulic safety net). We therefore calculated the percentage of grouped (non-solitary) vessels for different vessel size classes (RG_{VA}) as a fifth parameter.

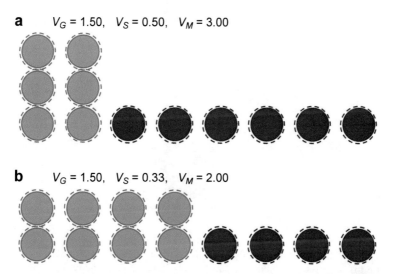

a $V_G = 1.50$, $V_S = 0.50$, $V_M = 3.00$

b $V_G = 1.50$, $V_S = 0.33$, $V_M = 2.00$

Figure 2. The same vessel grouping index V_G can be attained by a setup of differently large vessel groups V_M and a different vessel solitary fraction V_S. In the example of panel **a** with totally 12 vessels, 6 vessels belong to two groups of 3 vessels each, while 6 vessels are solitary. In the example of panel **b**, 8 vessels belong to four groups of 2 vessels, while only 4 vessels are solitary. The hydraulic integration in **b** is therefore greater than in **a**. Vessel lumen area is depicted by filled areas while the outlines of vessel walls are schematically represented by dashed lines.

Image analysis tool ROXAS

ROXAS is an image analysis tool for quantifying the xylem anatomy in cross-sectional samples of trees (angiosperms and conifers), shrubs and herbaceous plants. It is based on the image manipulation and registration capabilities of Image Pro Plus ≥ v.6.1 (Media Cybernetics, Silver Spring, Maryland, USA), and adds own code and algorithms to automatically detect conduits and, with some limitations, tree-ring boundaries (von Arx & Dietz 2005; Fonti *et al.* 2009; Wegner *et al.* 2013). The user can manually edit the automatically-generated results, and obtains a large set of output parameters for the entire sample including ring width, conduit lumen area, and vessel grouping parameters. ROXAS is, under certain conditions, free of charge for the research community (see www.wsl.ch/roxas for more details and a download link).

The identification of grouped *vs.* solitary vessels in the considered plane of the xylem cross section is based on distance criteria: to decide whether two neighboring vessels are touching each other, ROXAS first calculates the Euclidean distance d between their centroids and then subtracts the radius of each vessel to obtain the distance d' between the outlines of the vessel lumina (Fig. 3). The radius of the vessels is thereby adjusted for elliptical vessel shape. If d' is equal or smaller than the empirically determined threshold d_{thr} that approximately corresponds to the double-cell-wall thickness of the two neighboring vessels, they are considered as touching and hydraulically con-

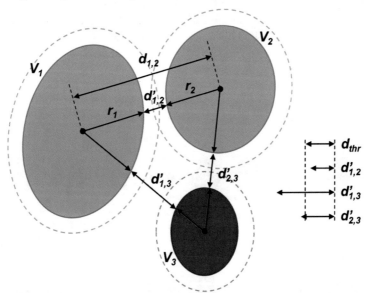

Figure 3. Conceptual model of how ROXAS distinguishes grouped from solitary vessels. The vessels are first expressed as their area-equivalent ellipses. Taking the example of vessels #1 (V_1) and #2 (V_2), the distance $d_{1,2}$ between the centroids of V_1 and V_2 is then calculated. From $d_{1,2}$, the radii r_1 and r_2 along the connection between the centroids is then subtracted to yield the distance $d'_{1,2}$ between the outlines of the two ellipses. If $d'_{1,2}$ is smaller than a threshold d_{thr}, which approximately corresponds to twice the thickness of average vessel walls, the two vessels are assumed to be grouped. Vessel lumen area is depicted by filled areas while the outlines of vessel walls are schematically represented by dashed lines.

nected through pits. This approach does not discriminate different intervessel contact length, *i.e.* the length of vessel wall in contact with a touching vessel as seen in the cross section (Jansen *et al.* 2011; Scholz *et al.* 2013); however, a longer contact length will indicate more intervessel pits and therefore potentially better hydraulic integration of the touching vessels.

RESULTS AND DISCUSSION

Plasticity of vessel grouping within species

Overall, 32,000 vessels were measured in the roots of the 14 individuals (mean number per individual: dry site: 3,800, moist site: 1,150). The individuals from the two sites differed significantly in the vessel grouping index (V_G), vessel solitary fraction (V_S), and the mean size of grouped vessels (V_M; Fig. 4). Individuals growing at the dry site had more grouped and fewer solitary vessels, and mean group size was also larger than in plants at the moist site. Similarly, the relative frequency of different group sizes (FD_G) also differed substantially, with plants growing at the dry site having larger maximum group sizes and a higher representation in all group sizes except the solitary ones (Fig. 5). Wider vessels were more often grouped than narrower vessels in the individuals at the dry site, while at the moist site, vessel grouping was independent from vessel size (Fig. 6). An example of how ROXAS detected vessel groups is shown in Figure 1c,d.

These results are consistent with other studies finding a positive relationship between vessel grouping and habitat dryness (Carlquist 1966; Robert *et al.* 2009; Lens *et al.* 2011; Carlquist 2012). Our results are also in accordance with a previous study by Robert *et al.* (2009) where considerable plasticity in vessel grouping patterns was found within species. Our findings suggest that a higher degree of grouping is both attained by fewer solitary vessels and larger groups of vessel multiples, which must not inevitably be linked (cf. Fig. 2). The dependence of

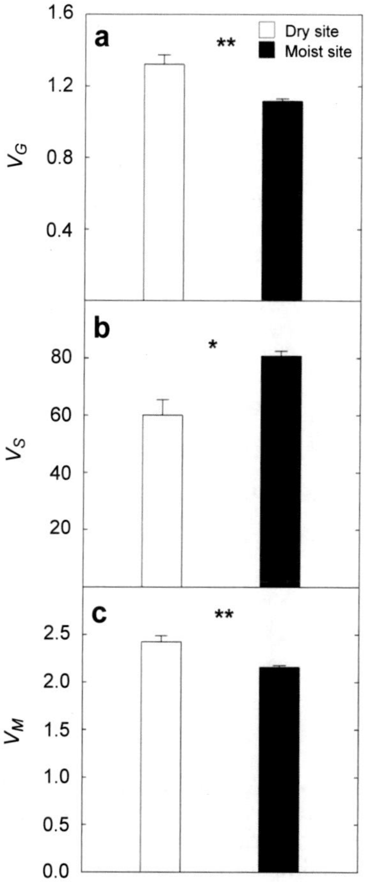

Figure 4. Differences in **a** vessel grouping index (V_G), **b** vessel solitary fraction (V_S), and **c** mean group size of non-solitary vessels (V_M) in *Verbascum thapsus* growing at a dry (n = 6) and moist site (n = 8) on Big Island, Hawaii, USA (mean ± 1 se). Statistical significance is based on t-tests, * = P ≤ 0.05, ** = P ≤ 0.01.

vessel grouping on vessel size is an intriguing result: based on our limited data set, the increased hydraulic integration of the widest vessels in dry-site plants could enhance pathway redundancy, if some of these vessels cavitate while other (smaller) vessels of the same vessel group remain functional.

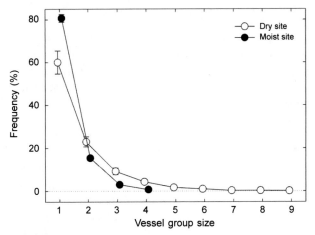

Figure 5. Relative frequency of different vessel group sizes (F_{DG}: including solitary vessels as groups of 1) in *Verbascum thapsus* growing at a dry and a moist site on Big Island, Hawaii, USA (mean ± 1 se). Vessel group sizes containing overall < 25 vessels (when multiplying a specific group size by the number of such groups) are not shown. Symbols within vessel group sizes are jittered for better readability.

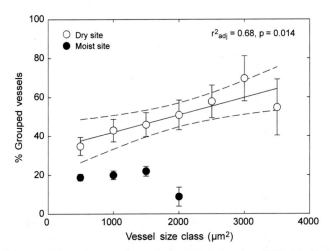

Figure 6. The percentage of grouped vessels per vessel size class (RG_{VA}) in *Verbascum thapsus* growing at a dry and a moist site on Big Island, Hawaii, USA (mean ± 1 se). Unlike at the moist site, grouping increased with vessel size in plants growing at the dry site. Only vessel size classes with ≥ 25 vessels included. Solid line – linear regression. Dashed line – 95 % confidence band.

Table 2. Comparison of vessel grouping parameters quantified with automatic ROXAS and manual inspection method for 3 individuals randomly chosen from the test dataset. See Table 1 for explanation of the acronyms of the vessel grouping parameters.

	Method	V_G	V_S	V_M	FD_G (number of vessels per group size)											
					1	2	3	4	5	6	7	8	9	10	11	12
Moist 1	ROXAS	1.096	0.837	2.155	1351	111	10	–	1	–	–	1	–	–	–	–
	Visual	1.096	0.851	2.191	1374	96	12	–	1	–	–	1	–	–	–	–
Moist 2	ROXAS	1.112	0.809	2.123	1121	109	46	–	–	–	–	–	–	–	–	–
	Visual	1.137	0.778	2.186	1192	128	27	1	–	–	–	–	–	–	–	–
Dry 1	ROXAS	1.434	0.504	2.563	2094	534	161	72	21	7	3	3	2	–	–	1
	Visual	1.412	0.526	2.602	2185	478	163	80	22	5	4	3	3	1	1	–

Accuracy of automatic detection of vessel grouping

The automatic results for V_G, V_S, and V_M differed from those obtained by manual inspection by 4% and less. (Table 2). The differences in the frequency distribution of different group sizes (FD_G) were, with a few exceptions, very small as well. The accuracy of the percentage of grouped vessels per vessel size class (RG_{VA}) could not be checked by manual inspection, but should behave similarly, as it is also based on detected vessel groups. This demonstrates that differences of the ROXAS from the manual inspection approach are negligible. This accordance depends on correct vessel detection and recognition of vessel outlines (avoiding under- and overestimation of vessel lumen area) and a sensible distance threshold d' for the distinction of touching $vs.$ non-touching vessels (cf. Materials and Methods). In addition, it was sometimes ambiguous by manual inspection, whether two closely positioned vessels were touching or not, which likely explains the few discrepancies to automatic results for FD_G.

Limitations of (automatic) vessel grouping analysis in cross sections

Two limitations require consideration when investigating vessel grouping in cross sections. First, when performing image analysis automatically or manually, a minimum cut-off vessel size has usually to be defined. This is because of limited image resolution, and because it is difficult to faithfully distinguish vessels from parenchyma cells below a certain size. If a very narrow vessel that is smaller than the threshold size connects to an otherwise solitary wider vessel, the latter will appear as solitary in the results (see Fig. 1b). As a result, vessel grouping will be underestimated. Yet, since this affects all samples, no bias should be expected, unless the narrowest vessels below the cut-off size are more frequent in some of the samples. Furthermore, from a hydraulic efficiency point of view, the narrowest vessels contribute very little to overall conductivity (Fonti *et al.* 2010), although they may be important as a "safety net".

Second, if the cut runs through oblique reticulate or scalariform perforation plates between two joining vessels or vessel elements (cf. Fig. 1b), they may mistakenly appear as paired vessels from a cross-sectional perspective (Carlquist 2001). This situation may occur quite frequently when considering vessel element lengths of a few hundred microns. However, simple perforation plates between vessel elements seem to be more abundant in hardwood species than scalariform or reticulate perforation plates (Butter-field & Meylan 1982). Accordingly, vessel elements in *V. thapsus* are connected through simple perforation plates, while only intervessel pits are scalariform (Schweingruber 2005). Vessel grouping may also be systematically overestimated if the cutting plane runs exactly through the branching of two vessels. This will be rather rare, however, because vessel branching does not occur extensively along the plant axis (André 2005); vessels rather connect through some axial contact length, which is correctly considered as grouping in this case. Some of these limitations could potentially be removed by 3D visualization techniques such as high-resolution computed tomography (Brodersen *et al.* 2011, 2013); however, accessibility and current limitations with respect to image resolution and sample size may pose challenges for more widespread application.

Perspectives for vessel grouping analysis

The presented example allows to anticipate the promises of future studies of vessel grouping. We demonstrated the value of characterizing vessel groupings – especially grouped versus solitary vessels – that is not possible through previously approaches of whole-sample analysis such as point pattern analysis (Loepfe *et al.* 2007; Mencuccini *et al.* 2010; Martínez-Vilalta *et al.* 2012). Yet, we expect that for a full understanding of structure-function relationships of the xylem hydraulic system further grouping indices than those proposed here will be needed. The additional parameters suggested in this paper serve as examples for moving beyond the well-established vessel grouping index V_G – which is a mean value with limited functional information – and characterizing also variation of vessel grouping. We hope that this paper stimulates innovation towards the most useful quantification of spatial vessel arrangement. Particularly, quantification of vessel size-dependent grouping may prove to better connect structural with functional xylem properties under environmental variability. It is well possible, for example, that quantifying vessel size-dependent grouping should extend to the question of whether two grouped vessels are of similar or different size. In this sense, automated image analysis as performed by ROXAS opens the door for new types of data analyses.

CONCLUSIONS

The image analysis tool ROXAS accurately calculated established and novel aspects of vessel grouping patterns. Thanks to its efficiency entire cross-sectional samples can be analyzed, which likely provides sufficiently large numbers of vessel measurements to detect relatively small differences within species and even among annual rings of individuals. The large number of vessels analyzed by ROXAS therefore constitutes an important methodological advancement. Other key advantages of automated quantification of vessel grouping lie within its objectivity and reproducibility.

Vessel grouping is a very poorly investigated functional trait of the xylem hydraulic system. Parameters used so far are rather simple and probably insufficient to cover the various functional meanings. Particularly unexplored parameters such as the frequency distribution of vessel group sizes, and the percentage of grouped (*vs.* solitary) vessels among different vessel size classes may prove to be of great ecological relevance and should therefore be considered in future studies. In addition, a better characterization of the structural context of vessels – such as the size of the vessels involved in grouping – may be necessary to understand whether vessel grouping is beneficial during water limitation. Thanks to its versatility, applications of ROXAS might also contribute to systematic studies that build on vessel grouping analyses (*e.g.* Lens *et al.* 2009), or to test hypotheses developed for woody clades with herbaceous species (*e.g.* Lens *et al.* 2013).

ACKNOWLEDGEMENTS

This work was partly supported by grants from the Swiss National Science Foundation (PBEZA-117266) to GvA, the Swiss State Secretariat for Education, Research and Innovation SERI (SBFI C12.0100) to PF and the USDA National Institute of Food and Agriculture Biology of Weedy and Invasive Species Program (NRI grant no. 2006-35320-17360) and ETH Zurich to CK. This study profited from discussions within the framework of the COST Action STReESS (COST-FP1106).

REFERENCES

Alder NN, Pockman WT, Sperry JS & Nuismer S. 1997. Use of centrifugal force in the study of xylem cavitation. J. Experim. Bot. 48: 665–674.

André J-P. 2005. Vascular organization of angiosperms: a new vision. Science Publishers Inc., Enfield, NH, USA.

Arbellay E, Fonti P & Stoffel M. 2012. Duration and extension of anatomical changes in wood structure after cambial injury. J. Experim. Bot. 63: 3271–3277.

Baas P & Carlquist S. 1985. A comparison of the ecological wood anatomy of the floras of Southern-California and Israel. IAWA Bull. n.s. 6: 349–353.

Baas P & Schweingruber FH. 1987. Ecological trends in the wood anatomy of trees, shrubs and climbers from Europe. IAWA Bull. n.s. 8: 245–274.

Baas P, Werker E & Fahn A. 1983. Some ecological trends in vessel characters. IAWA Bull. n.s. 4: 141–159.

Brodersen CR. 2013. Visualizing wood anatomy in three dimensions with high-resolution X-ray micro-tomography (µCT) – A review. IAWA J. 34: 408–424.

Brodersen CR, Lee EF, Choat B, Jansen S, Phillips RJ, Shackel KA, McElrone AJ & Matthews MA. 2011. Automated analysis of three-dimensional xylem networks using high-resolution computed tomography. New Phytol. 191: 1168–1179.

Brodersen CR, McElrone AJ, Choat B, Lee EF, Shackel KA & Matthews MA. 2013. In vivo visualizations of drought-induced embolism spread in *Vitis vinifera*. Plant Physiol. 161: 1820–1829.

Butterfield BG & Meylan BA. 1982. Cell wall hydrolysis in the tracheary elements of the secondary xylem. In: Baas P (ed.), New perspectives in wood anatomy: 71–84. Dr W. Junk Publishers, The Hague.

Cai J & Tyree MT. 2010. The impact of vessel size on vulnerability curves: data and models for within-species variability in saplings of aspen, *Populus tremuloides* Michx. Plant Cell Environm. 33: 1059–1069.

Carlquist S. 1966. Wood anatomy of Compositae: a summary, with comments on factors controlling wood evolution. Aliso 6: 25–44.

Carlquist S. 1984. Vessel grouping in dicotyledon wood: significance and relationship to imperforate tracheary elements. Aliso 10: 505–525.

Carlquist S. 2001. Comparative wood anatomy: systematic, ecological, and evolutionary aspects of dicotyledon wood. Springer, Berlin.

Carlquist S. 2009. Non-random vessel distribution in woods: patterns, modes, diversity, correlations. Aliso 27: 39–58.

Carlquist S. 2012. How wood evolves: a new synthesis. Botany 90: 901–940.

Fonti P, Eilmann B, Garcia-Gonzalez I & von Arx G. 2009. Expeditious building of ring-porous earlywood vessel chronologies without loosing signal information. Trees Structure Function 23: 665–671.

Fonti P, von Arx G, García-González I, Eilmann B, Sass-Klaassen U, Gärtner H & Eckstein D. 2010. Studying global change through investigation of the plastic responses of xylem anatomy in tree rings. New Phytol. 185: 42–53.

Hacke UG, Jacobsen AL & Pratt RB. 2009. Xylem function of arid-land shrubs from California, USA: an ecological and evolutionary analysis. Plant Cell Environm. 32: 1324–1333.

Hacke UG & Sperry JS. 2001. Functional and ecological xylem anatomy. Perspectives in Plant Ecol. Evol. Syst. 4: 97–115.

Höltta T, Vesala T, Peramaki M & Nikinmaa E. 2006. Refilling of embolised conduits as a consequence of 'Munch water' circulation. Funct. Plant Biol. 33: 949–959.

Jansen S, Gortan E, Lens F, Lo Gullo MA, Salleo S, Scholz A, Stein A, Trifilò P & Nardini A. 2011. Do quantitative vessel and pit characters account for ion-mediated changes in the hydraulic conductance of angiosperm xylem? New Phytol. 189: 218–228.

Kueffer C, Daehler CC, Torres-Santana CW, Lavergne C, Meyer JY, Otto R & Silva L. 2010. A global comparison of plant invasions on oceanic islands. Perspectives in Plant Ecol. Evol. Syst. 12: 145–161.

Lens F, Endress ME, Baas P, Jansen S & Smets E. 2009. Vessel grouping patterns in subfamilies Apocynoideae and Periplocoideae confirm phylogenetic value of wood structure within Apocynaceae. Amer. J. Bot. 96: 2168–2183.

Lens F, Sperry JS, Christman MA, Choat B, Rabaey D & Jansen S. 2011. Testing hypotheses that link wood anatomy to cavitation resistance and hydraulic conductivity in the genus *Acer*. New Phytol. 190: 709–723.

Lens F, Tixier A, Cochard H, Sperry JS, Jansen S & Herbette S. 2013. Embolism resistance as a key mechanism to understand adaptive plant strategies. Curr. Opin. in Plant Biol. 16: 287–292.

Loepfe L, Martinez-Vilalta J, Pinol J & Mencuccini M. 2007. The relevance of xylem network structure for plant hydraulic efficiency and safety. J. Theor. Biol. 247: 788–803.

Lopez BC, Sabatae S, Gracia CA & Rodriguez R. 2005. Wood anatomy, description of annual rings, and responses to ENSO events of *Prosopis pallida* HBK, a wide-spread woody plant of arid and semi-arid lands of Latin America. J. Arid Environm. 61: 541–554.

Martínez-Vilalta J, Mencuccini M, Álvarez X, Camacho J, Loepfe L & Piñol J. 2012. Spatial distribution and packing of xylem conduits. Amer. J. Bot. 99: 1189–1196.

Mencuccini M, Martinez-Vilalta J, Piñol J, Loepfe L, Burnat M, Alvarez X, Camacho J & Gil D. 2010. A quantitative and statistically robust method for the determination of xylem conduit spatial distribution. Amer. J. Bot. 97: 1247–1259.

Nardini A, Dimasi F, Klepsch M & Jansen S. 2012. Ion-mediated enhancement of xylem hydraulic conductivity in four *Acer* species: relationships with ecological and anatomical features. Tree Physiol. 32: 1434–1441.

Parker IM, Rodriguez J & Loik ME. 2003. An evolutionary approach to understanding the biology of invasions: Local adaptation and general-purpose genotypes in the weed *Verbascum thapsus*. Conserv. Biol. 17: 59–72.

Robert EMR, Koedam N, Beeckman H & Schmitz N. 2009. A safe hydraulic architecture as wood anatomical explanation for the difference in distribution of the mangroves *Avicennia* and *Rhizophora*. Funct. Ecol. 23: 649–657.

Sano Y, Morris H, Shimada H, Ronse De Craene LP & Jansen S. 2011. Anatomical features associated with water transport in imperforate tracheary elements of vessel-bearing angiosperms. Ann. Bot. 107: 953–964.

Scholz A, Klepsch M, Karimi Z & Jansen S. 2013. How to quantify conduits in wood? Frontiers in Plant Science 4.

Schweingruber FH. 2005. The xylem database. Birmensdorf. http://www.wsl.ch/dendro/xylemdb [accessed 17 June 2013].

Seipel T, Alexander JM, Daehler CC, Edwards PJ, Dar PA, McDougall K, Naylor B, Parks C, Reshi ZA, Rew LJ, Schroder M & Kueffer C. 2013. Home and away: performance of an invasive plant in native and non-native regions. J. Biogeogr. In revision.

Sperry JS & Tyree MT. 1988. Mechanism of water stress-induced xylem embolism. Plant Physiol. 88: 581–587.

Tyree MT, Davis SD & Cochard H. 1994. Biophysical perspectives of xylem evolution: is there a tradeoff of hydraulic efficiency for vulnerability to dysfunction? IAWA J. 15: 335–360.

Tyree MT & Ewers FW. 1991. The hydraulic architecture of trees and other woody plants. New Phytol. 119: 345–360.

von Arx G, Archer SR & Hughes MK. 2012. Long-term functional plasticity in plant hydraulic architecture in response to supplemental moisture. Ann. Bot. 109: 1091–1100.

von Arx G & Dietz H. 2005. Automated image analysis of annual rings in the roots of perennial forbs. Int. J. Plant Sci. 166: 723–732.

von Arx G & Dietz H. 2006. Growth rings in the roots of temperate forbs are robust annual markers. Plant Biol. 8: 224–233.

Wegner L, von Arx G, Sass-Klaassen U & Eilmann B. 2013. ROXAS - an efficient and accurate tool to detect vessels in diffuse-porous species. IAWA J. 34: 425–432.

Wheeler JK, Sperry JS, Hacke UG & Hoang N. 2005. Inter-vessel pitting and cavitation in woody Rosaceae and other vesselled plants: a basis for a safety versus efficiency trade-off in xylem transport. Plant Cell Environm. 28: 800–812.

Zimmermann MH. 1983. Xylem structure and the ascent of sap. Springer, Berlin.

Accepted: 6 September 2013

IAWA Journal 34 (4), 2013: 446–458

BRILL

FIRE INFLUENCE ON *PINUS HALEPENSIS*: WOOD RESPONSES CLOSE AND FAR FROM THE SCARS

Veronica De Micco[1,*], **Enrica Zalloni**[1], **Angela Balzano**[1], and **Giovanna Battipaglia**[2]

[1]Dipartimento di Agraria, Università degli Studi di Napoli Federico II, via Università 100,
I-80055 Portici (NA), Italy
[2]Dipartimento di Scienze e Tecnologie Ambientali, Biologiche e Farmaceutiche (Di.S.T.A.Bi.F),
Seconda Università di Napoli, via Vivaldi 43, I-81100 Caserta, Italy
*Corresponding author; e-mail: demicco@unina.it

ABSTRACT

Tree rings provide information about environmental change through recording stress events, such as fires, that can affect their growth. The aim of this study was to investigate wood growth reactions in *Pinus halepensis* Mill. trees subjected to wildfires, by analysing anatomical traits and carbon and oxygen isotope composition. The study area was Southern France where two sites were selected: one subjected to fires in the last 50 years, the other characterised by comparable environmental conditions although not affected by fire events (control site). We analysed whether wood growth depends on the tangential distance between developing xylem cells and the limit where the cambium was directly damaged by fire. In the burnt site, thick wood sections, including fire-scar, were taken from surviving plants. Digital photo-micrographs were analysed to measure early- and latewood width, wood density, and tracheid size. Anatomical and isotopic traits were analysed in two series of tree rings (5 rings before and 5 after the fire) selected at different positions along the circumference (close or far from the scar). Anatomical and isotopic traits were quantified also on tree rings of the same years from cored trees growing at the control site. Results showed different wood reaction tendencies depending on the distance from the scar. The comparison between plants from the two sites allowed to exclude possible climate interference.

Our results are discussed in terms of two kinds of growth reactions: the local need to promptly compartmentalise the scarred cambial zone and sapwood after fire, and the general growth perturbations due to tree reaction to crown scorch during fire. Anatomical results, combined with dendrochronological and isotopic analysis, could provide an efficient way to distinguish between direct growth reactions due to heat-related damage on cambium and indirect outcomes related to defoliation.

Keywords: Aleppo pine, fire, quantitative wood anatomy, scar, stable isotopes, tree rings.

DOI 10.1163/22941932-00000036

INTRODUCTION

Tree rings record information about environmental factors affecting plant growth. In Mediterranean-type ecosystems, wild- and man-induced fires have played a key role in regulating plant growth and survival, priming different fire-response strategies, thus resulting in specific ecosystem dynamics and a peculiar fire-shaped landscape made of mosaic communities (Barbero *et al.* 1987; Di Pasquale *et al.* 2004; Thompson 2005). Plant responses to fires can be variable even within a given species depending on fire intensity, growth season when the event occurs and plant age (Trabaud 1981). Wildfires are considered among natural hazards which are responsible for tree injury; such fire-induced damages determine growth anomalies in tree-ring series which are commonly analysed within dendrogeomorphic research based on the "process-event-response" concept (Schröder 1978; Stoffel *et al.* 2010). Fire damage can determine changes in tree growth due to both crown destruction and trunk injuries, commonly known as fire scars.

If a fire event is not destructive and trees survive, plants experience either a temporary growth reduction or a growth increase (Schweingruber 1996). The temporary decrease in tree-ring growth is a response typically ascribed to crown scorch and leaf/needle surface reduction, while temporary growth increase can be due to other phenomena such as reduced inter-plant competition, increased release of nutrients or better light conditions (Brown & Swetnam 1994; Schweingruber 1996, 2007).

From an anatomical viewpoint, after a fire event, cambium can be destroyed and the formation of new cells can be interrupted in the injured sector of the trunk or branch. The reaction of plants to cambium destruction due to fire has generally been compared with common responses after mechanical wounding. Trees react to cambium destruction after mechanical wounding through subsequent mechanisms: wood compartmentalisation, the formation of callus tissue and cell overgrowth at the edges of the injury in order to reduce the exposed area up to the complete closure of the wound (Shigo 1984; Schweingruber 2007). Depending on the species and the period of the year when wounding occurs, traumatic resin ducts can be produced at different times following injury (Stoffel *et al.* 2010). To study tree ring's reaction to fires, it would be desirable to analyse the whole trunk section. Since this is not always feasible, coring techniques are often applied. It has been highlighted how the position of cores is fundamental to avoid gaining misleading information (Stoffel *et al.* 2010). However, studies aiming to analyse the reaction of tree-ring growth at anatomical level by systematically comparing the variability of responses in samples taken at different distances from scars are not frequent. Recently, Bigio *et al.* (2012) demonstrated that the combination of information about the position of scars and the measurement of other anatomical parameters is needed for accurate intra-annual dating of fire events in *Castanea sativa* Mill. The authors also hypothesise that specific anatomical traits, observed in subsamples along the circumference of analysed branches, arise in response to local heating of the cambium and likely to canopy damage.

Wood reaction to fire-related injury is not only a mechanical response but can also be primed by physiological alterations due to direct and indirect effects of burning. Such physiological reactions have been investigated also by combining growth observations with the analysis of stable isotopes in tree rings formed before and after fire events (Beghin *et al.* 2011). Carbon and oxygen isotopes measured in the same tree rings have proved to be a useful tool for understanding the response of trees to both climate and environmental parameters (Ehleringer *et al.* 1993) especially related to stress events (Moreno-Gutiérrez *et al.* 2012; Battipaglia *et al.* 2013a). Indeed, δ^{13}C and δ^{18}O are indicators for both temperature and moisture regime changes: fractionation processes during CO_2 uptake are important for δ^{13}C, while changes in the soil and leaf water isotope ratio are determining factors for δ^{18}O (McCarroll & Loader 2004). Thus, simultaneous measurement of δ^{18}O and δ^{13}C in tree rings allows discrimination between biochemical and stomatal limitations to photosynthesis (Scheidegger *et al.* 2000) during stress events, such as fire.

More detailed understanding about how trees react to wildfires may be obtained through a combined approach, involving quantitative wood anatomy and stable isotope analysis, applied to dendrochronology. Applying different disciplines to study the effect of environmental factors on tree-ring growth has proved to be a valuable approach to gain a more comprehensive understanding of wood physiological responses (De Micco *et al.* 2007, 2012; Battipaglia *et al.* 2010, 2013b; Novak *et al.* 2013; Rosner 2013).

In this paper, we analyse wood growth reactions in *Pinus halepensis* Mill. subjected to a wildfire with the aim to: a) verify whether they depend on the tangential distance between developing xylem cells and the limit where the cambium was directly damaged by fire (scar edge), and b) to hypothesise possible mechanisms triggering eventual differential responses. To reach these aims, anatomical and isotopic traits of tree rings before and after fire were analysed in two radii along the circumference (close and far from the fire-scar edge). More specifically, we aimed to test the hypotheses that: a) isotopic composition of cell walls should be independent from the sampling distance to the scar, being an indicator of the physiological status of the whole tree, and b) anatomical traits should be more variable and might be considered as indicators of wood growth reactions primed by different biological demands or constraints depending on the tangential distance from the scar edge.

MATERIALS AND METHODS

Study site and plant material

The two sampling sites were located in Cotignac, Southern France (43° 33' N, 6° 07' E, 316–350 m asl). They were comparable in their environmental and physiographic conditions (slope, elevation, exposition) but they represented a different fire history. The first was subjected to two fires in the last 50 years: the first occurred in 1979, the second in 1999. The second site was never affected by fires and was used as control, not-burnt site. Fire history for all stands were based on the Promethee database (http://www.promethee.com) that collects information about fire-events in Southern France since 1973.

The climate is typically Mediterranean with hot and dry summers (annual drought), promoting fire events. Monthly mean temperature and precipitation data (1957–2011) were obtained from the Cotignac meteorological station, 20 km from the sampling site.

The mean annual temperature ranged from 8.6 ± 5.4 (Tmin) to $17.5 \pm 6.2\,°C$ (Tmax); the mean temperature of the hottest season (summer, June to August, JJA) ranged between 15.2 ± 1.2 (Tmin) and $25.2 \pm 3.1\,°C$ (Tmax), while the mean temperature of the coldest season (winter, December to February) was from 2.1 ± 0.5 (Tmin) to $10.3 \pm 0.6\,°C$ (Tmax). The total annual precipitation was 752 ± 127 mm with 105 ± 103 mm in the driest season (JJA). In the two sites, vegetation was dominated by a *Pinus halepensis* Mill. plantation with an understory of *Quercus coccifera* L., *Phillyrea* spp., *Rosmarinus officinalis* L. and *Thymus vulgaris* L.

In each site, five trees of *Pinus halepensis* were selected. From 10 trees growing at the burnt site and showing fire-scars, both thick stem sections in the region of the scar and cores opposite to the region of trunk injury were taken (Fig. 1). Thick sections were sampled with a chainsaw, while cores were sampled with an increment borer (Haglöf, Långsele, Sweden) of 0.5 cm in diameter. From trees growing at the control sites, only cores were sampled. Cores were carefully mounted on channelled wood; both thick cross sections and cores were seasoned in a fresh-air dry store and sanded a few weeks later. The anatomical and isotopic variations due to the fire event of 1979 were analysed.

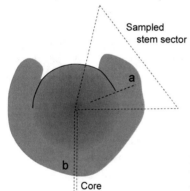

Figure 1. Scheme representing a scarred trunk section with positions where sub-samples were taken: a, close to scar (CS position) and b, far from scar (FS position).

Microscopy and digital image analysis

Both thick stem sections and cores were observed under a reflected light microscope (BX60, Olympus, Hamburg, Germany). Tree rings were identified and visually dated. In trees growing at the burnt site, the region close to the scar was analysed under the microscope to analyse wood traits arisen in response to burning. Moreover, the 5 years preceding and the 5 years formed after the 1979 fire event were selected at two radii along the circumference: close (Fig. 1, position a, CS-close to scar) and far from the fire-scar edge (Fig. 1, position b, FS-far from scar). The position CS was established as the wood region just adjacent to the limit where the scar ends: at this limit, signs of burnt wood, of tracheid occlusion and browning are not evident anymore. The position FS was just opposite to the middle of scar width. In cores extracted from trees growing at control sites, tree rings corresponding to 1974–1978 and 1980–1984 were selected as well. Microphotographs of each selected tree ring were obtained with a digital camera (CAMEDIA C4040, Olympus). Digital images were analysed with the software program

AnalySIS 12.0 (Olympus) in order to measure: tree-ring width, early- and latewood width, density (percent of wood occupied by cell walls), and tracheid lumen area.

Carbon and oxygen stable isotopes

δ^{13}C- and δ^{18}O-values were determined on the same rings selected for anatomical analyses: the 5 rings preceding and the 5 rings formed after the fire event, with sub-samples taken in the CS and FS positions along the circumference (Fig. 1, positions a and b). The sub-samples were milled for cellulose extraction according to the method described by Loader *et al.* (1997) and modified by Battipaglia *et al.* (2008).

δ^{13}C values were determined using an elemental analyser linked to an isotopic ratio mass spectrometer (MS, Finnigan Delta S) via a variable open split interface (Conflo II, all from Finnigan Mat, Bremen). δ^{18}O was measured using a continuous-flow pyrolysis system, with an elemental analyser (Carlo Erba 1108, Italy) linked to an isotope ratio mass spectrometer (MS, Delta Plus XP, ThermoFinnigan, Germany) (Saurer *et al.* 1998).

Isotopic compositions are reported as delta values (δ) in ‰ 'units' relative to an internationally accepted reference: Vienna PeeDee belemnite (VPDB) for carbon isotope values and Vienna Standard Mean Ocean Water (VSMOW) for oxygen isotope values. The standard deviation for the repeated analysis of an internal standard (commercial cellulose) was better than 0.1‰ for carbon and better than 0.3‰ for oxygen.

Data provided by Francey *et al.* (1999) and McCarroll and Loader (2004) were used to remove the decline in the δ^{13}C of atmospheric CO_2 due to fossil fuel emissions from the carbon isotope data series. The corrected series were then employed in all the statistical analyses.

Statistical analyses

All results were subjected to statistical analysis (ANOVA) using the SPSS statistical package (SPSS Inc., Chicago, IL, USA). Data on wood density were transformed through arcsine function before statistical analysis.

RESULTS

The analysed trunks of *Pinus halepensis* reacted to fire through the typical reaction of wood to wounding. The two scars related to the two fire events were evident in the thick trunk sections (Fig. 2). The scar region is characterised by burnt wood (Fig. 3): in the burnt area, xylem cell walls appeared completely black. Wood reacted to fire wounding also by separating the wounded zone from the healthy part of xylem through the formation of a barrier zone. At the scar level, moving towards inner tracheid layers, the phenomenon of cell wall browning was accompanied by tracheid occlusion, suggesting the occurrence of chemical compartmentalisation. Moreover, a callus tissue was formed (Fig. 4) and subsequent regeneration of the vascular tissue at the wound edge occurred, thus determining wood overgrowth at the scar edge in order to favour the wound closure. Compartmentalisation was also evidenced by the formation of several layers of traumatic resin ducts (Fig. 5).

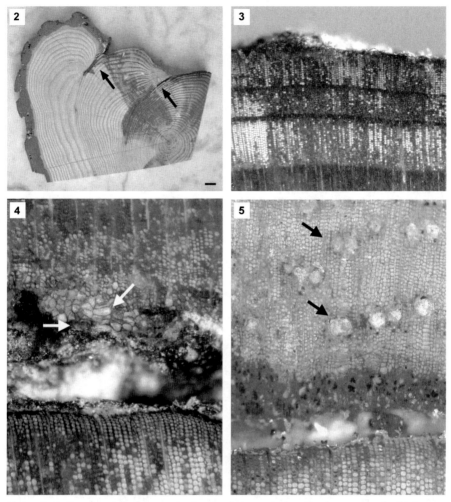

Figure 2–5. Cross sections of a trunk of *Pinus halepensis* showing fire scars. – **2**: Thick cross sections sampled at the scar level (black arrows); light microscopy views of cross sections showing; **3**: burnt wood and occluded tracheids with browning cell walls; **4**: callus tissue (white arrows); **5**: traumatic resin ducts (black arrows). — Scale bar = 1 cm.

The quantification of anatomical features showed that there were no significant differences in any of the considered parameters between plants growing at the fire and control site before the burning event (data not shown). Thus, possible micro-climate interferences were excluded and only data from trees at the burnt site are reported.

Tree-ring width was not significantly different between the two sampled positions along the trunk circumference (CS-close to scar and FS-far from scar) before the fire event (Fig. 6). In both positions incidence of latewood was similar: latewood width represented $19.0 \pm 0.821\%$ of the total ring width at the CF position, while it was $20.3 \pm 0.993\%$ at FS. Conversely, after burning, ring width was significantly lower far

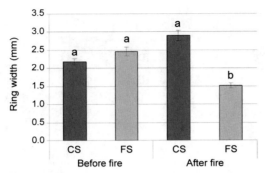

Figure 6. Comparison of rings developed close to (CS) and far from (FS) the scar in both ring series formed before and after fire, on the basis of ring width. Mean values and standard errors are shown. Different letters correspond to significantly different values within periods of wood formation (*i.e.* before and after fire).

from the scar than close to wounding (Fig. 6); moreover, rings far from the scar showed a significantly higher incidence of latewood ($21.8 \pm 1.25\%$) than those formed close to the scar ($16.1 \pm 0.851\%$).

In both earlywood and latewood, intra-wood variability of density (expressed as percent of space occupied by cell walls) was high even before the fire event and determined

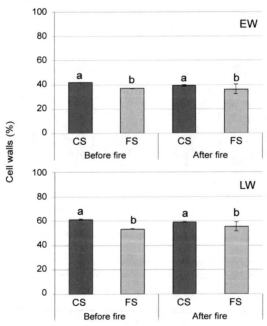

Figure 7. Comparison of rings developed close to (CS) and far from (FS) the scar in both ring series formed before and after fire, on the basis of wood density (expressed as percent of volume occupied by cell walls), in earlywood (EW) and latewood (LW). Mean values and standard errors are shown. Different letters correspond to significantly different values within periods of wood formation (*i.e.* before and after fire).

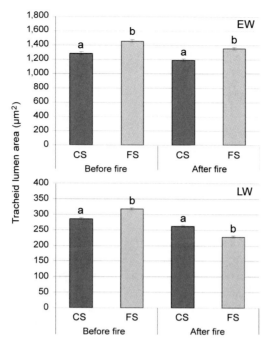

Figure 8. Comparison of rings developed close to (CS) and far from (FS) the scar in both ring series formed before and after fire, on the basis of tracheid lumen area in earlywood (EW) and latewood (LW). Mean values and standard errors are shown. Different letters correspond to significantly different values within periods of wood formation (*i.e.* before and after fire).

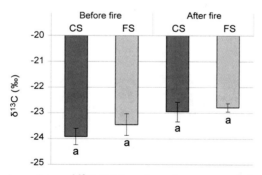

Figure 9. Comparison of mean δ¹³C (‰) tree-ring values measured close to (CS) and far from (FS) the scar in both ring series formed before and after fire. Mean values and standard errors are shown. Different letters correspond to significantly different values within periods of wood formation (*i.e.* before and after fire).

the occurrence of significant differences between the two positions CS and FS (Fig. 7). In particular, earlywood and latewood of trees located far from the scar presented a decrease in cell wall proportion of 11.8 % and 12.6 %, respectively, compared to that measured close to the scar. After the fire event, this trend of variation was maintained: wood at FS position showed a significantly lower cell wall percentage than in the

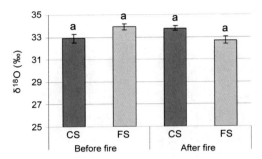

Figure 10. Comparison of mean $\delta^{18}O$ (‰) tree-ring values measured close to (CS) and far from (FS) the scar in both ring series formed before and after fire. Mean values and standard errors are shown. Different letters correspond to significantly different values within periods of wood formation (*i.e.* before and after fire).

CS zone (Fig. 7). More specifically, cell wall proportion far from the scar decreased, respectively, by 7.9% and 5.9% compared to that measured close to the scar.

In earlywood, variation trends for tracheid size between the two positions were similar before and after fire: at FS position, tracheid lumen area was significantly higher than at CS (Fig. 8, EW). In contrast, in latewood, opposite trends were found between the two positions before and after fire indicating a fire-driven reaction: rings formed after burning showed narrower tracheids far from the scar than close to wounded area (Fig. 8, LW).

As regards stable isotopes, $\delta^{13}C$ and $\delta^{18}O$ ratios did not change significantly between the two analysed positions along the trunk circumference either before or after the fire event (Fig. 9 & 10). However, the tendency of variation of $\delta^{18}O$ was inverted before and after wounding (Fig. 10).

DISCUSSION

Wood reacts to fire injuries, as to other wounding due to biotic or abiotic factors, by altered, sometimes enhanced, xylem growth to isolate the wounded area with a barrier zone as quickly as possible (Shigo 1984; Schweingruber 2007). The analysis of the microsections of *Pinus halepensis* close to the scar contained the typical signs of chemical compartmentalisation, accompanied by the formation of a callus tissue, which promotes the overgrowth of new xylem tissue, and of traumatic resin ducts. Cells of the callus, as well as of tracheids formed before and after the fire event, were filled with occluding materials deposited by the adjoining ray cells. Indeed, cell proliferation in the callus tissues is needed to regenerate new xylem through the redifferentiation and regeneration of lateral meristems at the wound edge (Larson 1994). Moreover, lignification and suberisation of cell walls at the callus level, as well as accumulation of phenolic compounds, appearing as brown deposits occluding cell lumens, are reported as defence reactions to prevent pathogen attack (Larson 1994; Pearce 1996). Moving towards newly formed cell layers, callus cells develop into woundwood which is characterised by short and reorienting tracheids often interrupted by additional resin canals (Oven & Torelli 1999). Moving even more outwardly, after the reorientation zone, normal tracheid formation is restored. While the analysis of anatomical reactions in the scar zone can furnish information about growth response after direct damage to the cambial zone, the characterisation of anatomical and isotopic composition of wood moving farther from the scar edge can be useful to gain other information more related to the direct effect of cambium heating or to the indirect effect of crown reduction (Schweingruber 2007; Bigio *et al.* 2012). We showed that, in *P. halepensis,* both carbon and oxygen isotopes composition were not dependent on the tangential distance between growing xylem cells and the limit where the cambium was directly damaged by fire. This means that, since the difference of isotope composition between tree rings sampled close or far to the scar was not significant either before or after burning, trends of $\delta^{13}C$ and $\delta^{18}O$ variations were linked to the natural intra-annual variability along the trunk radius. In contrast, different anatomical traits showed different reaction tendencies depending on the tangential distance from the scar. Moreover, such tendencies were different in earlywood and latewood. After fire, wider rings with a lower latewood proportion characterised by larger tracheids were formed close to the scar compared with rings far from the scar. These anatomical features might be interpreted as a sort of direct low-cost mechanical reaction. We hypothesise that the local formation of wider rings would be primed by the local need to promptly protect the scar after fire; however, a lower latewood fraction of larger tracheids would allow the formation of a large wood volume, but requiring less resources for cell wall construction soon after the stress event. Indeed, after cambial cell division, the second and third steps of xylem element differentiation, namely the formation of a thick multi-layered secondary cell wall and lignin deposition, are complex and metabolite-demanding processes (Plomion *et al.* 2001). However, to support our hypothesis, studies on the chemical composition of cell walls are needed.

Conversely, the variation of anatomical traits far from the scar would more likely be an indicator of physiological effects. The formation of narrower rings with higher latewood proportion characterised by narrower tracheids would indicate a generally lowered tree growth, probably due to defoliation after-fire. The occurrence of the same trends of variation in wood density and earlywood tracheid size between the two sampling positions in both rings formed before and after the fire event indicates that such anatomical parameters are not good indicators of wood growth perturbations due to fire. This is not in agreement with the general statement that cell wall thickening and increased lignification occur as typical reactions to defoliation and pollarding (Schweingruber 2007).

It is recognised that, although anatomical parameters would be very useful to understand the effect of fire on tree growth, anatomical methods are rarely coupled with dendrochronological studies aiming to study forest reaction to burning (Bigio *et al.* 2012). This is the reason why wood response to fire is generally compared with wood reaction to other types of stresses such as mechanical wounding, pathogen attack (Kajii *et al.* 2013), natural or simulated defoliation, floods or other extreme weather conditions which can alter cambium functioning and subsequent phenomena of cell differentiation. These wounding conditions are generally associated with physiological stresses in trees whose common response is reduced radial growth accompanied by reduced conduit dimensions, either of tracheids or vessels (Asshoff *et al.* 1998; St. George *et al.* 2002; Corcuera *et al.* 2004; Jones *et al.* 2004).

In conclusion, while for carbon and oxygen isotopes tendencies of variations in tree rings formed before and after fire do not depend on the sub-sampling position along the trunk circumference, anatomical data need to be kept separated between positions along the trunk circumference in order to avoid misleading results. The lack of significant differences in isotope composition between sampling positions are probably due to the fact that isotopes are good proxies of a mean physiological response of plants to stress events. In fact, the $\delta^{13}C$ and the $\delta^{18}O$ are tightly related to change in photosynthetic activity and/or stomatal conductance (Farquhar *et al.* 1989) that are influenced by fire in the whole trees. This suggests that isotope analysis is not sensitive to distinguish between mechanical and physiological responses after wounding. In contrast, different responses of the variation of anatomical properties to burning along the trunk circumference can be interpreted as primed by either mechanical and biological demands (need to promptly close the wounded zone) or physiological constraints (*e.g.* crown reduction). In conclusion, anatomical results combined with dendrochronological data could provide an efficient way to distinguish between direct growth reactions due to heat-related damage of the cambium and indirect outcomes related to other stressors.

ACKNOWLEDGEMENTS

The authors thank Thomas Fournier for sharing wood sections sampled during his PhD project "Structure des communautés et diversité spécifique post-incendie en forêt méditerranéenne" supervised by Cristopher Carcaillet. The authors also acknowledge Giovanna Aronne (University of Naples Federico II) for sharing her microscopy laboratory and offering her expertise during the development of this research.

REFERENCES

Asshoff R, Schweingruber FH & Wermelinger B. 1998. Influence of a Gypsy Moth (*Lymantria dispar* L.) outbreak on radial growth and wood-anatomy of Spanish Chestnut (*Castanea sativa* Mill.) in Ticino (Switzerland). Dendrochronology 16–17: 133–145.

Barbero M, Bonin G, Loisel R, Miglioretti F & Quézel P. 1987. Impact of forest fires on structure and architecture of Mediterranean ecosystems. Ecologia Méditerranea 13: 39–50.

Battipaglia G, De Micco V, Brand WA, Linke P, Aronne G, Saurer M & Cherubini P. 2010. Variations of vessel diameter and $\delta^{13}C$ in false rings of *Arbutus unedo* L. reflect different environmental conditions. New Phytol. 188: 1099–1112.

Battipaglia G, De Micco V, Brand WA, Saurer M, Aronne G, Linke P & Cherubini P. 2013b. Drought impact on water use efficiency and intra-annual density fluctuations in *Erica arborea* on Elba (Italy). Plant Cell Environm. In press, doi: 10.1111/ pce.12160.

Battipaglia G, Jäggi M, Saurer M, Siegwolf RTW & Cotrufo MF. 2008. Climatic sensitivity of $\delta^{18}O$ in the wood and cellulose of tree rings: results from a mixed stand of *Acer pseudoplatanus* L. and *Fagus sylvatica*. Palaeogeogr. Palaeoclim. 261: 193–202.

Battipaglia G, Saurer M, Cherubini P, Calfapietra C, McCarthy HR, Norby RJ & Cotrufo MF. 2013a. Elevated CO_2 increases tree-level intrinsic water use efficiency: insights from carbon and oxygen isotope analyses in tree rings across three forest FACE sites. New Phytol. 197: 544–554.

Beghin R, Cherubini P, Battipaglia G, Siegwolf R, Saurer M & Bovio G. 2011. Tree-ring growth and stable isotopes (^{13}C and ^{15}N) detect effects of wildfires on tree physiological processes in *Pinus sylvestris* L. Trees 25: 627–636.

Bigio E, Gärtner H & Conedera M. 2012. Fire-related features of wood anatomy in a sweet chestnut (*Castanea sativa*) coppice in southern Switzerland. Trees 24: 643–655.

Brown PM & Swetnam TW. 1994. A cross-dated fire history from coast redwood near Redwood National Park, California. Can. J. For. Res. 24: 21–31.

Corcuera L, Camarero JJ & Gil-Pelegrin E. 2004. Effects of a severe drought on growth and wood anatomical properties of *Quercus faginea*. IAWA J. 25: 185–204.

De Micco V, Battipaglia G, Brand WA, Linke P, Saurer M, Aronne G & Cherubini P. 2012. Discrete versus continuous analysis of anatomical and δ^{13}C variability in tree rings with intra-annual density fluctuations. Trees 26: 513–524.

De Micco V, Saurer M, Aronne G, Tognetti R & Cherubini P. 2007. Variations of wood anatomy and δ^{13}C within tree rings of coastal *Pinus pinaster* Ait. showing intra-annual density fluctuations. IAWA J. 28: 61–74.

Di Pasquale G, Di Martino P & Mazzoleni S. 2004. Forest history in the Mediterranean region. In: Mazzoleni S, Di Pasquale G, Mulligan M, Di Martino P & Rego F (eds.), Recent dynamics of the Mediterranean vegetation and landscape: 13–20. Wiley & Sons Ltd, Chichester.

Ehleringer JR, Hall AE & Farquhar GD. 1993. Stable isotopes and plant carbon-water relations. Academic Press, Inc., San Diego.

Farquhar GD, Ehleringer JR & Hubick KT. 1989. Carbon isotope discrimination and photosynthesis. Annu. Rev. Plant Mol. Physiol. Plant Mol. Biol. 40: 503–537.

Francey RJ, Allison CE, Etheridge DM, Trudinger CM, Enting IG, Leuenberger M, Langenfelds RL, Michel E & Steele LP. 1999. A 1000-year high precision record of δ^{13}C in atmospheric CO_2. Tellus. Series B, Chem. Phys. Meteorol. 51: 170–193.

Jones B, Tardif JW & Westwood E. 2004. Weekly xylem production in trembling aspen (*Populus tremuloides*) in response to artificial defoliation. Can. J. Bot. 82: 590–597.

Kajii C, Morita T, Jikumaru S, Kajimura H, Yamaoka Y & Kuroda K. 2013. Xylem dysfunction in *Ficus carica* infected with wilt fungus *Ceratocystis ficoila* and the role of the vector beetle *Euwallacea interjectus*. IAWA J. 34: 301–312.

Larson PR. 1994. The vascular cambium: development and structure. Springer Verlag, Berlin.

Loader NJ, Robertson I, Barker AC, Switsur VR & J Waterhouse JS. 1997. An improved technique for the batch processing of small whole wood samples to α-cellulose. Chem. Geol. 136: 313–317.

McCarroll D & Loader NJ. 2004. Stable isotopes in tree rings. Quat. Sci. Rev. 23: 771–801.

Moreno-Gutiérrez C, Battipaglia G, Cherubini P, Saurer M, Nicolás E, Contreras S & Querejeta JI. 2012. Stand structure modulates the long-term vulnerability of *Pinus halepensis* to climatic drought in a semiarid Mediterranean ecosystem. Plant Cell Environm. 35: 1026–1039.

Novak K, Saz Sánchez MA, Čufar K, Raventós J & de Luis M. 2013. Age, climate and intra-annual density fluctuations in *Pinus halepensis* in Spain. IAWA J. 34: 459–474.

Oven P & Torelli N. 1999. Response of the cambial zone in conifers to wounding. Phyton 39: 133–137.

Pearce RB. 1996. Tansley Review No. 87. Antimicrobial defences in the wood of living trees. New Phytol. 132: 203–233.

Plomion C, Leprovost G & Stokes A. 2001. Wood formation in trees. Plant Physiol. 127: 1513–1523.

Rosner S. 2013. Hydraulic and biomechanical optimization in Norway spruce trunkwood – A review. IAWA J. 34: 365–390.

Saurer M, Robertson I, Siegwolf R & Leuenberger M. 1998. Oxygen isotope analysis of cellulose: an interlaboratory comparison. Anal. Chem. 70: 2074–2080.

Scheidegger Y, Saurer M, Bahn M & Siegwolf R. 2000. Linking stable oxygen and carbon isotopes with stomatal conductance and photosynthetic capacity: a conceptual model. Oecologia 125: 350–357.

Schröder JF. 1978. Dendrogeomorphological analysis of mass movement on Table Cliffs Plateau, Utah. Quat. Res. 9: 168–185.

Schweingruber FH. 1996. Wood structure and environment. Springer-Verlag, Berlin, Heidelberg.

Schweingruber FH. 2007. Tree rings and environment. Dendroecology. Paul Haupt Publishers, Bern, Stuttgart, Vienna.

Shigo AL. 1984. Compartmentalization – A conceptual framework for understanding how trees grow and defend themselves. Ann. Rev. Phytopatol. 22: 189–214.

St. George S, Nielson E, Conciatori F & Tardif J. 2002. Trends in *Quercus macrocarpa* vessel areas and their implications for treering paleoflood studies. Tree-Ring. Res. 58: 3–10.

Stoffel M, Bollschweiler M, Butler DR & Luckman BH. 2010. Tree rings and natural hazards: an introduction. In: Stoffel M, Bollschweiler M, Butler DR & Luckman BH (eds.), Tree rings and natural hazards: A state-of-the-art: 3–23. Springer, Dordrecht, Heidelberg, London, New York.

Thompson JD. 2005. Plant evolution in the Mediterranean. Oxfor University Press, Oxford.

Trabaud L. 1981. Man and fire: impacts on Mediterranean vegetation. In: di Castri F, Goodall DW & Specht RL (eds.), Ecosystems of the world 11. Mediterranean-type shrublands: 523–537. Elsevier Scientific Publishing Company, Amsterdam.

Accepted: 26 August 2013

IAWA Journal 34 (4), 2013: 459–474

BRILL

AGE, CLIMATE AND INTRA-ANNUAL DENSITY FLUCTUATIONS IN *PINUS HALEPENSIS* IN SPAIN

Klemen Novak[1,2,3,*], **Miguel Angel Saz Sánchez**[1], **Katarina Čufar**[2], **Josep Raventós**[3] and **Martin de Luis**[1]

[1]University of Zaragoza, Department of Geography and Regional Planning, C/Pedro Cerbuna 12, 50009 Zaragoza, Spain
[2]University of Ljubljana, Biotechnical Faculty, Department of Wood Science and Technology, Rožna dolina, Cesta VIII/34, 1000 Ljubljana, Slovenia
[3]University of Alicante, Department of Ecology, Carretera San Vicente del Raspeig s/n, 03690 San Vicente del Raspeig-Alicante, Spain
*Corresponding author: e-mail: kn4@alu.ua.es

ABSTRACT

Intra-annual density fluctuations (IADFs) in tree rings of Aleppo pine (*Pinus halepensis*) are considered to be among the most promising wood anatomical features in dendrochronological studies. They provide environmental information in addition to those obtained from tree-ring widths. We used a network of 35 sites in Spain, ranging from nearly desert to temperate climate. We analysed tree-ring series of 529 trees to study IADF frequencies, and their dependence on climatic factors and cambial age. The results showed that IADF frequency is age dependent, with its maximum at the cambial age of 27 years (evaluated at breast height). The frequencies varied across the network and at different sites we recorded that 0.3 % to 33 % of the analysed tree rings contained IADFs. They were more frequent where and when the temperatures were higher, summer drought was intense and autumn was the main precipitation season. IADF formation was particularly related to high minimum temperatures and wet conditions in late summer and autumn. These results suggest that IADF formation is not related to stressful conditions during summer but to favourable conditions during autumn. These conditions promote cambial reactivation and consequently formation of wider tree rings.

Keywords: Aleppo pine, wood structure, tree rings, Mediterranean.

INTRODUCTION

The Aleppo pine (*Pinus halepensis* Mill.) is an important and widespread tree species in the Mediterranean and can grow under widely diverse climatic conditions (Barbéro *et al.* 1998; Richardson & Rundel 1998; De Micco *et al.* 2013) on a great variety of substrates and on poor soil. It is thermophilous and heliophilous, and tolerant to high temperatures and drought, but does not cope well with excessive humidity, frost and snow (Girard *et al.* 2012). Due to its growth plasticity and adaptability to different site and climatic conditions, it is an important species to study the effect of climatic change

© International Association of Wood Anatomists, 2013
Published by Koninklijke Brill NV, Leiden

DOI 10.1163/22941932-00000037

on trees across the Mediterranean. In this area, the summer drought represents the main constraint for tree growth with great inter-annual variations in duration and intensity (Girard *et al*. 2012). Climatic models predict progressive warming and reduction of precipitation (Christensen *et al*. 2007) which are expected to endanger the survival of trees, especially at more extreme sites (Alcamo *et al*. 2007).

In this context, dendrochronology enables us to study past responses of trees to climate (Nicault *et al*. 2008) and helps us to predict future vegetation shifts in response to climatic change. *Pinus halepensis* has a typical conifer wood structure, containing resin canals and clearly distinguishable tree rings with earlywood (EW) and latewood (LW), and more or less gradual transition between them (Schweingruber 1988). However, deviations from such normal structure are frequent and are characterised by abrupt changes in ring width, variable frequency of normal and of traumatic resin canals, and intra-annual density fluctuations (IADFs) (De Luis *et al*. 2007; Novak *et al*. 2011; Olivar *et al*. 2012).

The combined approach of dendrochronology and quantitative wood anatomy has been also used to characterise IADFs in dated tree rings. Tree rings containing IADFs can be in some cases divided into different types, like the ones with latewood-like tracheids within the earlywood (E-rings), or with earlywood-like tracheids within the latewood (L-ring) (Campelo *et al*. 2007a). A recent wood formation study in *P. halepensis* on an extremely dry site in Spain has shown that L-type IADFs are formed as a consequence of cambial reactivation in autumn after its stop or slowdown during hot and dry summers (De Luis *et al*. 2011b). However, another recent study in *P. halepensis* of various sites in Spain has demonstrated that L-rings are much more frequent than E-rings (Novak *et al*. 2013).

Dendrochronological studies of *P. halepensis* proved that IADFs provide valuable information of climate-growth relationship in addition to the information obtained from tree-ring widths (Novak *et al*. 2013). This is in agreement with other reports on the high importance of density, wood structural features, and cell dimensions in dendrochronology and ecological studies (Battipaglia *et al*. 2010, 2013; Fonti *et al*. 2010; Martin-Benito *et al*. 2013; Campelo *et al*. 2013; Panayotov *et al*. 2013). However, the information obtained from IADFs may greatly vary depending on site conditions, population structure and inter-annual as well as intra-annual variability in environmental conditions of the sites (Novak *et al*. 2013). As a consequence, our knowledge of the main climatic factors promoting IADFs across different environmental conditions is still deficient.

The main purpose of this study was to use a dense and diverse dendroclimatic network in Spain to establish the frequency variation of IADFs in *P. halepensis*, and to determine the climatic factors which promote IADF formation, as well as whether the processes are age dependent.

MATERIAL AND METHODS

Sampling sites and climatic conditions

The study was carried out in the Mediterranean area in Spain, at 35 different sites with different climatic conditions (Fig. 1; Table 1). Mean annual temperatures on the sites ranged from 10.9 °C (the coldest site) to 18.3 °C (the warmest site) and mean

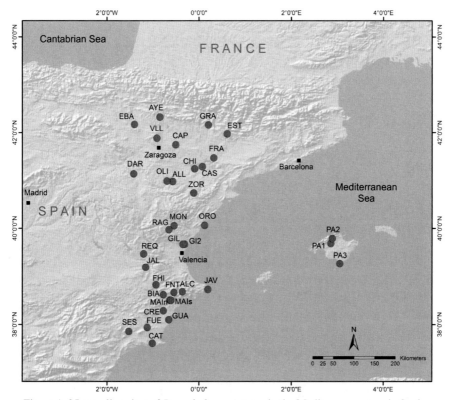

Figure 1. 35 sampling sites of *Pinus halepensis* trees in the Mediterranean area in Spain.

annual precipitation from 243 mm (the driest site) to 1181 mm (the wettest site). The distribution of total precipitation is from 6 to 22 % in summer and from 27 to 44 % in autumn.

Sampling was conducted on mature, apparently healthy Aleppo pine (*Pinus halepensis* Mill.) trees, without any visible damage. We selected 9 to 33 trees per site and extracted 1 to 8 cores per tree at breast height, altogether 1054 samples from 529 trees in total were studied.

The cores were then labelled, fixed on wooden supports, air-dried and sanded with progressively finer grades of sandpaper (80, 180, 300, 500 grit) until the tree-ring structure was clearly visible under the stereo microscope.

Tree-ring measurements and standardisation procedures

The tree-ring series were visually and statistically crossdated and compared between each other by calculating the t-value after Baillie and Pilcher (1973) using the TSAP-Win Scientific program (version 4.68e). Additionally the quality of crossdating was verified using the COFECHA program - version 6.06P (Holmes 1994).

The tree-ring widths (TRW) were measured under a stereo microscope with an accuracy of 0.01 mm, with the TSAP-Win Scientific program and LINTAB™5

Table 1 – *For legends, also see the next page*

Site	Code	Chron. time span	Altitude (m)	Nr. trees	Nr. samples	T max (°)	T min (°)	P (mm)	winter P (%)	spring P (%)	summer P (%)	autumn P (%)
Alcoy-Font Roya	FNT	1864-2006	1022	14	24	20.1	10.1	352	0.26	0.29	0.10	0.35
Alcoy-Penaguila	ALC	1852-2001	674	15	30	19.9	8.4	377	0.26	0.30	0.11	0.33
Alcubierre	CAP	1800-2006	738	14	27	19.0	7.8	371	0.21	0.29	0.22	0.28
Alloza	ALL	1888-2006	595	17	31	20.3	7.6	380	0.19	0.30	0.22	0.29
Ayerbe	AYE	1946-2006	924	19	33	16.9	4.8	701	0.28	0.26	0.16	0.30
Biar	BIA	1927-2000	806	15	29	20.6	7.3	321	0.23	0.30	0.15	0.33
Cartagena	CAT	1915-2007	116	15	27	22.7	12.5	267	0.31	0.26	0.06	0.37
Caspe	CAS	1845-2007	166	15	28	21.7	9.3	241	0.20	0.30	0.17	0.33
Chiprana	CHI	1900-2003	160	7	9	22.3	9.6	300	0.20	0.28	0.19	0.33
Crevillente	CRE	1832-2000	285	12	43	23.5	13.2	272	0.25	0.27	0.10	0.38
Daroca	DAR	1934-2006	937	14	28	17.8	6.3	425	0.19	0.31	0.23	0.27
Ejea-Bardenas	EBA	1909-2003	365	7	7	19.2	7.3	399	0.23	0.29	0.19	0.30
El Grado	GRA	1946-2006	168	15	30	19.0	5.4	567	0.22	0.29	0.19	0.30
Estopiñan del Castillo	EST	1964-2006	502	14	27	19.4	5.7	482	0.20	0.29	0.21	0.31
Font de la Figuera	FHI	1946-2011	680	14	28	20.0	7.4	335	0.22	0.31	0.16	0.31
Fraga	FRA	1844-2006	340	14	29	21.9	9.2	304	0.20	0.30	0.16	0.34
Fuensanta	FUE	1902-2007	138	14	26	24.5	11.2	274	0.25	0.30	0.10	0.34
Gilet-P44	GI2	1899-2011	140	15	30	21.8	12.3	274	0.23	0.24	0.11	0.42
Gilet-Sancti Spiritu	GIL	1892-2006	175	14	27	22.0	13.3	428	0.23	0.23	0.10	0.43
Guardamar	GUA	1912-2006	10	13	26	21.4	14.6	274	0.29	0.25	0.07	0.40
Jalance	JAL	1863-2004	571	22	49	20.2	7.7	385	0.26	0.29	0.14	0.32
Javea	JAV	1944-2000	96	15	60	21.0	15.2	589	0.28	0.21	0.08	0.43
Maigmo-norte	MAI	1867-2009	845	15	30	19.7	8.3	377	0.26	0.30	0.11	0.33
Maigmo-sur	MAI	1901-2009	762	25	50	22.4	11.5	256	0.24	0.29	0.11	0.37
Mallorca-Binissalem	PA1	1914-2009	120	13	26	21.7	12.6	407	0.29	0.22	0.09	0.39
Mallorca-Caimari	PA2	1888-2009	386	13	26	19.4	7.7	1181	0.33	0.23	0.06	0.38
Mallorca-Cap Salines	PA3	1890-2009	14	12	24	21.7	14.5	299	0.28	0.20	0.08	0.44
Montanejos	MON	1955-2001	569	16	31	20.3	8.4	489	0.23	0.24	0.16	0.37
Oliete	OLI	1977-2006	530	15	28	19.8	7.2	329	0.18	0.31	0.21	0.29
Oropesa	ORP	1921-2003	1	15	30	23.6	11.7	459	0.23	0.24	0.13	0.41
Puerto de Ragudo	RAG	1965-2011	959	15	30	19.0	7.1	501	0.21	0.28	0.17	0.34
Requena	REQ	1789-2003	721	15	31	20.8	7.6	390	0.25	0.27	0.15	0.33
Sierra Espuña	SES	1894-2007	846	16	29	21.6	10.4	289	0.24	0.33	0.09	0.34
Villanueva de Gállego	VLL	1878-2006	452	15	29	19.7	8.7	335	0.21	0.30	0.19	0.30
Zorita	ZOR	1832-2001	857	15	30	19.4	8.2	462	0.21	0.27	0.20	0.32

Average annual climatic conditions for the period 1951–2007

Table 1. Description of 35 sampling sites of *Pinus halepensis* trees in the Mediterranean area in Spain: site name and code, chronology time span, altitude, number of trees and samples, and average annual climatic conditions for the period 1951–2007: mean maximal annual temperatures (T_{max}), mean minimal annual temperatures (T_{min}), mean annual precipitation (P) and seasonal distribution of precipitation in winter (% winter P), spring (% spring P), summer (% summer P) and autumn (% autumn P).

←

measuring device (RINNTECH e.K., Hardtstrasse 20-22, D-69124 Heidelberg, Germany, www.rinntech.com).

Quantitative wood anatomy was used to characterise intra-annual density fluctuations (IADFs) in dated tree rings. We analysed 81,238 tree rings with a stereo microscope. Since in many cases it was not possible to objectively differentiate between different IADF types, we only classified their presence in the tree rings. We assigned the value 1 if the IADF was present in the individually dated tree rings, and the value 0 if it was not observed (Fig. 2). Only one person completed the classification, in order to obtain comparable results.

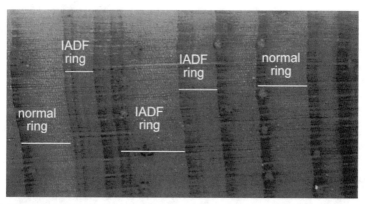

Figure 2. Tree rings of *Pinus halepensis*: normal rings, and rings showing intra-annual density fluctuations (IADFs).

Raw series of TRW and of IADF frequencies were calculated individually for each of the 529 analysed trees using the arithmetic mean of the available samples.

To study age-related growth trends, the raw series of IADF frequencies were aligned by biological cambial age observed at breast height considering pith-offset estimations, and averaged for each age using the arithmetic means. Then, a three-parameter Weibull function (Equation 1) was employed to explore the relationship between the cambial age and the average IADF frequency observed for each age.

$$IADF_{(f)} = a \cdot b \cdot c \cdot Age^2 \cdot e^{-a \cdot Age^b} \qquad \text{Equation 1}$$

Where $IADF_{(f)}$ is the observed frequency of tree-rings containing IADF, Age is the cambial age observed at breast height of the trees and a, b and c are fitting parameters.

The resulting Weibull equation was later used as a regional curve for detrending purposes (Briffa *et al.* 1992; Esper *et al.* 2003). Thus, for each individual tree, the standardised frequency series of IADFs ($IADF_{(Std\ f)}$) were calculated as the difference between the observed and the predicted $IADF_{(f)}$.

The arithmetic means of $IADF_{(Std\ f)}$ series, which are age independent, were then calculated for each study site to describe spatial variations in the occurrence of IADFs across the network.

The climate–IADF relationship

Monthly values of total precipitation, average maximum and minimum temperature collected at each site in the period from 1950 to 2007 were obtained from Spain02 database (Herrera *et al.* 2012).

The information from 27,430 tree rings for which the climatic data were available was used to identify the climatic conditions that promote and trigger the formation of IADFs.

For each analysed tree ring, we calculated different climatic parameters (maximal temperatures, minimal temperatures and total precipitation) for months between September of the previous year to November of the current year and for the climatic seasons: winter (December, January, February), spring (March, April, May), summer (June, July, August) and autumn (September, October, November). For each analysed climatic parameter we divided the observed range of values into the percentiles. Then, for each percentile class, we calculated the average of the standardised IADF frequencies ($IADF_{(Std\ f)}$) as observed in individual tree rings. Pearson's Product Moment Correlation Coefficient was used to measure strength of association between the average of the $IADF_{(Std\ f)}$ and the average value of the percentile class.

Differences in widths between the tree rings containing IADFs and those without IADFs

For each set of site/year conditions, the ratio between the average ring width calculated for the tree rings containing IADFs and the average ring width calculated for tree rings without IADFs was computed. Then, for each study site, we applied a T-test between the obtained series of ratios in order to establish the differences in ring width between tree rings containing IADFs and tree rings without IADFs.

RESULTS

Tree-ring network and effect of age on IADF frequency

The established tree-ring network consists of 529 tree-ring series of *Pinus halepensis* from 35 sites in Spain. The average length of tree-ring series is 81 years, ranging from 24 to 215 years. Of the 81,238 analysed tree rings, 6,937 (8.54%) showed IADFs. The frequency of IADFs varied with cambial age and reached its maximum of 10.5% at the cambial age at breast height of 27 years. With the three-parameter Weibull function we confirmed that IADF frequency depends on cambial age ($r^2 = 0.874$; $p < 0.01$) as shown in Figure 3.

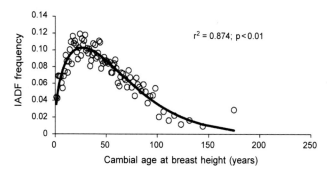

Figure 3. The effect of cambial age (horizontal axis) on IADF frequency (vertical axis) as explained with three-parameter Weibull function ($IADF_{(f)}$ is the observed frequency of tree-rings containing IADF, Age is the cambial age observed at the breast height of the trees and a, b and c are fitting parameters).

Table 2. Description of IADF frequency (raw frequency ($IADF_{(f)}$ and standardised frequency $IADF_{(std\,f)}$) and ratio between tree-ring widths with IADF and tree-ring widths without IADF ($TRW_{(IADF)} / TRW_{(no\,IADF)}$, average and p value) at 35 sampling sites.

| Site name | Code | IADF frequency | | $TRW_{(IADF)} / TRW_{(no\,IADF)}$ | |
		$IADF_{(f)}$	$IADF_{(std\,f)}$	Average	p value
Alcoy-Penaguila	ALC	0.0083	-0.0544	1.3886	0.0247
Alcoy-Font Roja	FNT	0.0386	-0.0361	1.3331	0.0110
Alcubierre	CAP	0.0119	-0.0667	0.9927	0.9239
Alloza	ALL	0.0363	-0.0402	1.3762	0.0003
Ayerbe	AYE	0.0228	-0.0673	1.3822	0.0799
Biar	BIA	0.1190	0.0306	1.0620	0.1243
Cartagena	CAT	0.0788	-0.0037	1.7765	0.0006
Caspe	CAS	0.0091	-0.0554	1.1860	0.3839
Chiprana	CHI	0.0383	-0.0394	1.6676	0.0014
Crevillente	CRE	0.0574	-0.0190	1.3933	0.0000
Daroca	DAR	0.1751	0.0861	1.3268	0.0004
Ejea-Bardenas	EBA	0.0210	-0.0665	0.8796	0.2620
El Grado	GRA	0.0033	-0.0870	1.2032	0.3991
Estopiñan del Castillo	EST	0.0601	-0.0307	1.5839	0.0041
Font de la Figuera	FHI	0.0870	-0.0029	1.2073	0.0643
Fraga	FRA	0.0164	-0.0445	1.4410	0.0079
Fuensanta	FUE	0.2526	0.1726	1.5799	0.0000
Gilet-Sancti Spiritu	GIL	0.1600	0.0806	1.5725	0.0021
Gilet-P44	GI2	0.1718	0.0920	1.6654	0.0000
Guardamar	GUA	0.1580	0.0750	1.3544	0.0000
Jalance	JAL	0.0591	-0.0143	1.2350	0.0016
Javea	JAV	0.3300	0.2422	1.3162	0.0000
Maigmo-norte	MAIn	0.0494	-0.0282	1.4956	0.0000
Maigmo-sur	MAIs	0.0837	0.0029	1.0558	0.3179
Mallorca-Binissalem	PA1	0.1957	0.1163	1.7457	0.0000
Mallorca-Caimari	PA2	0.1580	0.0775	1.6831	0.0000
Mallorca-Cap Salines	PA3	0.0657	-0.0069	1.8026	0.0001
Montanejos	MON	0.0461	-0.0446	0.9667	0.5743
Oliete	OLI	0.1237	0.0328	1.0302	0.7076
Oropesa	ORO	0.1635	0.0773	1.2853	0.0016
Puerto de Ragudo	RAG	0.0598	-0.0312	1.1799	0.0595
Requena	REQ	0.0233	-0.0312	1.6106	0.0000
Sierra España	SES	0.0812	0.0049	1.3456	0.0000
Villanueva de Gállego	VLL	0.0165	-0.0582	1.1062	0.3154
Zorita	ZOR	0.0407	-0.0309	1.2757	0.0026

Spatial variation in the frequency of IADFs

The frequency of IADFs varies across the geographical distribution of *Pinus hale-pensis* in Spain (Fig. 4; Table 2). The lowest frequency of 0.3 % (IADF$_{(Std f)}$ = -0.09) was observed in Ayerbe, close to the Pyrenees, at an altitude of 924 m; the highest frequency of 33 % (IADF$_{(Std f)}$ = 0.24) was observed in Javea on the Mediterranean coast, at an altitude of 96 m. The occurrence of IADFs is generally more frequent on the sites near the coast and on the Balearic Islands (Mallorca). IADFs are generally less frequent in the inland, in the mountains and especially on the northern part of the distribution range of *P. halepensis*.

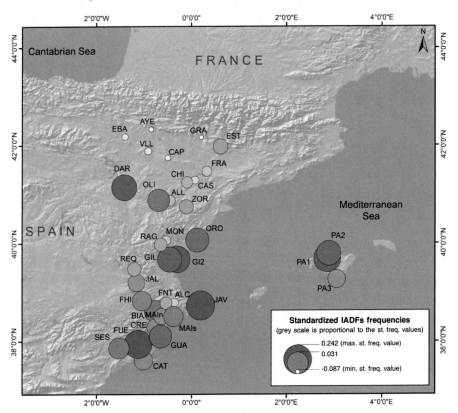

Figure 4. Spatial variation in the occurrence of standardized IADF frequencies, from the lowest (smallest point) of 0.3 % (*e.g.* Ayerbe) to the highest frequency (largest point) with 33 % (*e.g.* Javea).

The climate – IADF relationship

Analysis of spatio-temporal variations of IADFs showed that they are more frequent at the sites and in years with warm climatic conditions, especially in terms of minimal temperatures, and where/when autumn is the main precipitation season (Fig. 5 & 6). Hot and dry summers were hypothesised to be the most prominent and stressful climatic element promoting IADF formation. Interestingly, the IADFs are not related to

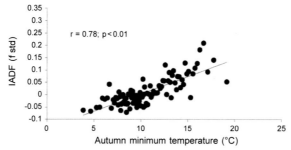

Figure 5. Increasing standardized IADF frequencies with increasing autumn minimum temperature. The correlations are significant.

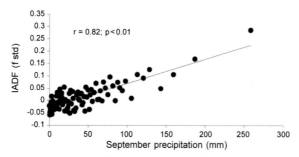

Figure 6. Increasing standardized IADF frequencies with increasing September precipitation. The correlations are significant.

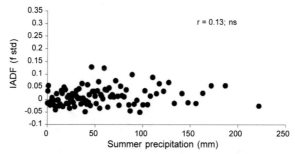

Figure 7. Standardized IADF frequencies and summer maximum temperature. The correlations are not significant.

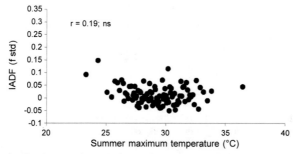

Figure 8. Standardized IADF frequencies and summer precipitation. The correlations are not significant.

summer conditions in terms of maximal summer temperatures and total amount of precipitation (Fig. 7 & 8); they are strongly related to autumn conditions. Higher minimum temperatures and higher precipitation in autumn, especially in September have proved to be the most critical climatic elements promoting IADF formation. This suggests that favourable conditions for cambial production – which are probably related to its reactivation after summer drought – may trigger IADF formation.

Differences between the widths of tree rings containing IADFs and the widths of tree rings without IADFs

Generally, IADFs occur in wider tree rings; the tree rings containing IADFs are on average 1.42 times wider than those without IADFs (Table 2). Statistical differences between the widths of the tree rings containing IADFs and the widths of tree rings without IADFs were found in 23 of the 35 analysed sites (p < 0.05). Higher differences were observed for the Balearic sites (Island of Mallorca) and coastal areas. For 12 of the sites analysed, mainly inland, no differences in tree ring widths were found (Fig. 9).

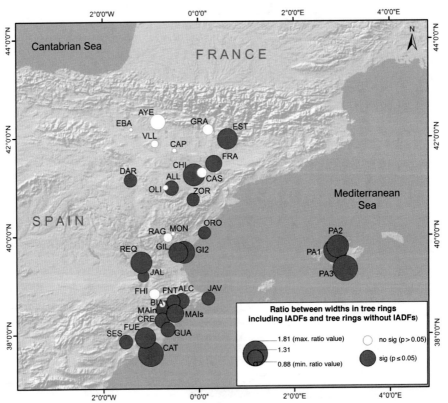

Figure 9. Mean ratio between the widths of tree rings containig IADFs and the widths of tree rings without IADFs on 35 sites. The circles marked in dark gray indicate the sites where the tree rings containing IADFs are significantly wider than the tree rings without IADFs. The white circles represent the sites without significant differences between the width of tree rings with and without IADFs.

DISCUSSION

The study of anatomical features of wood represents a promising approach for a better interpretation of the influence of seasonal climate on tree rings (Fonti *et al.* 2010; von Arx *et al.* 2013; Wegner *et al.* 2013). IADFs in tree rings can be used as a key intra-annual feature, because of their relation to climatic variability and change, and their occurrence in numerous species growing in different environmental conditions, *e.g.* *Pinus halepensis* (De Luis *et al.* 2011a; Olivar *et al.* 2012), *Pinus pinea* (Campelo *et al.* 2007a), *Pinus pinaster* (De Micco *et al.* 2007; Vieira *et al.* 2009; Rozas *et al.* 2011; Campelo *et al.* 2013), *Arbutus unedo* (Battipaglia *et al.* 2010, 2013; De Micco *et al.* 2012), *Quercus ilex* (Campelo *et al.* 2007b), *Juniperus virginiana* (Edmondson 2010), *Pinus banksiana* (Copenheaver *et al.* 2006; Hoffer & Tardif 2009), *Pinus syl-vestris* (Rigling *et al.* 2001; Panayotov *et al.* 2013), *Pinus elliottii* var. *densa* (Harley *et al.* 2012), *Picea abies* (Zubizarreta-Gerendiain *et al.* 2012), or *Pseudotsuga menziesii* (Martinez-Meier *et al.* 2008).

In addition, the climatic information that can be obtained from IADFs is comple-mentary to the climatic signals obtained from tree-ring widths (Novak *et al.* 2013), and the occurrence of IADFs is connected with intra-annual dynamics of wood formation (De Luis *et al.* 2007, 2011b).

The study of IADFs in wood has a great potential, but there are still problems in their analysis. First, different studies use different classifications for IADFs (E-rings, L-rings, early-IADF, mid-IADF, late-IADF), depending on the position of IADFs within the tree ring. The type and position of IADF is normally visually determined by examining the tree rings under a stereomicroscope, and IADFs are dated and assign-ed to earlywood or latewood subjectively (Rigling *et al.* 2001; Campelo *et al.* 2007a; Battipaglia *et al.* 2010; De Micco *et al.* 2012). In our previous research of *P. halepen-sis* growing in contrasted environmental conditions, we found that a pre-cise classifica-tion of IADFs, despite being subjective, can be performed (Novak *et al.* 2013).

However, at transitional sites which were included in the present study, such classi-fication is even more subjective. This is the main reason that in this study of trees from a great variety of sites we adopted a more conservative approach of just classifying the presence or absence of IADFs in the tree rings.

In addition, IADF frequencies (as well as other biological parameters like tree-ring widths) are known to be age/size dependent, which has been previously demonstrated by comparing the frequencies in the trees of different ages/sizes (De Luis *et al.* 2009; Vieira *et al.* 2009; Campelo *et al.* 2013). However, it has yet to be established whether this relation is linear or not. In the presented dendrochronological network of *P. hale-pensis* in Spain we demonstrated that the relationship between IADF frequency and age exists. The IADF frequency, observed in samples taken at breast height, increases with cambial age for the first 27 years and then decreases. We explained the effect of the age with a three-parameter Weibull function, which proved to be a useful tool for a description of the effect of age on IADF frequency, and as a model for detrending and interpretation of the effects of climate on IADF formation. We propose to use and test this model in other species and environmental conditions.

Our approach also shows limitations related to tree age, which was determined on the samples extracted from the trees at breast height, and does not represent exact tree age. It is difficult to resolve this problem, because the majority of dendrochronological studies are based on samples extracted at breast height. In future studies it would be interesting to contrast different effects, like tree height, tree size, and age, to improve the Weibull model as a detrending tool.

Comparison of IADF frequency and climatic influences across the distribution of the species is difficult due to differences in population structure (age). In this context, the Weibull detrending model, as a common procedure used across the network, represents an important advance in making the sites comparable. Our study demonstrated that IADF frequency varies across the geographical distribution of *P. halepensis*, and that the occurrence of IADFs is distributed across a clear geographical/environmental gradient. IADFs are more frequent at sites and in the years with warm climatic conditions, and where/when autumn is the main precipitation season (mainly at coastal sites). In contrast, in colder and dryer conditions in autumn, IADFs are scarce (inland or on high elevated sites).

The studies relating IADFs and climate are mainly based on information from local sites and there are few studies reporting how climatic influences vary across environmental gradients (Campelo *et al.* 2007a; Battipaglia *et al.* 2010; Olivar *et al.* 2012; Cherubini *et al.* 2013; Mamet & Kershaw 2013; Novak *et al.* 2013). In the current study we used combined spatial and temporal variations in IADF frequency across a wide geographical range including a wide environmental gradient in a single analysis. From the statistical point of view, the methodology is quite simple and we propose that it should be tested in other species and/or environmental conditions. In our analysis we found, contrary to expectations and variations described for other species, that summer conditions (especially maximal and minimal temperatures, as well as precipitation) did not explain spatial and temporal variations of IADF frequency (Copenheaver *et al.* 2010; Olivar *et al.* 2012; Zubizarreta-Gerendiain *et al.* 2012; Campelo *et al.* 2013). Our results are partly in agreement with the results of Vieira *et al.* (2010) who found that IADF frequency in latewood of *Pinus pinaster* was positively related to autumn precipitation, and with Campelo *et al.* (2007a), who observed the same in *Pinus pinea*. It should also be noted that *P. halepensis* in Spain predominately shows L-type IADFs with earlywood-like cells in latewood which are formed after reactivation of cambium in autumn (De Luis *et al.* 2011b; Novak *et al.* 2013).

In our study, occurrence of IADFs proved to be related to suitable conditions for growth in autumn with mild temperatures (mainly minimum temperatures) and suitable wetness. The results are in agreement with our previous studies, which have demonstrated that IADF frequencies in adult (De Luis *et al.* 2011b) and juvenile trees of *P. halepensis* (De Luis *et al.* 2011a) are related to cambial reactivation after summer. According to this interpretation, IADF frequencies are not directly related to summer stress, but to favourable conditions in autumn, which promote cambial reactivation.

Our results also demonstrate that tree rings containing IADFs are wider than those without IADFs. This is in line with the former interpretation since the reactivation in autumn permits the trees to complete a second period of radial growth. Campelo *et al.* (2013) and Copenheaver *et al.* (2006) also report that tree rings with IADFs are wider, which suggests that IADF formation is related to favourable growing conditions. In contrast, Bogino and Bravo (2009) have shown radial growth of *Pinus pinaster* subsp. *mesogeensis* to be negatively correlated with the presence of IADFs. A possible explanation could be that *P. halepensis* is an extraordinarily plastic species, able of reactivation after summer if the conditions are favourable. Other species perhaps do not have such a plastic character and thus no reactivation. In addition, the predominant environmental conditions in our study areas (Western Mediterranean) are quite special within the Mediterranean area, and are characterised by regularly abundant autumn precipitation. It should be interesting to explore whether *P. halepensis* growing in the Eastern Mediterranean, where the precipitation is mainly restricted to winter, also presents the same growth pattern, a similar frequency of IADFs in the wood and a similar response to climatic factors promoting their formation.

CONCLUSIONS

IADF frequency in *Pinus halepensis* proved to be age dependent and showed an asymmetric bell-shaped distribution with its maximum at the cambial age of 27 years at breast height.

The effect of age on IADF frequency can be explained with a three-parameter Weibull function, which proved to be a useful tool, both for description as well as a model for detrending and interpretation of the effects of climate on IADF formation.

IADF frequency varies across the geographical distribution of *P. halepensis*, with a clear geographical/environmental gradient. IADFs are more frequent at the sites and in the years with warm climatic conditions, and where/when autumn is the main precipitation season (coastal sites). In contrast, under colder and dryer conditions in autumn, the presence of IADFs is scarce (inland or high elevated sites).

Spatio-temporal analysis revealed that IADF formation is strongly related to warm conditions (especially with minimum temperatures) in summer, dry conditions in late spring and summer, and wet conditions in late summer and autumn. This indicates that IADF formation is not related to stressful conditions during summer, but to favourable conditions during autumn which promote cambial reactivation.

These results suggest that IADF formation indicates plasticity of *P. halepensis* and its ability to resume cambial activity after summer drought.

Our results also show that tree rings containing IADFs are wider than those without IADFs, suggesting that IADF formation is not related to stressful but rather to favourable climatic conditions.

ACKNOWLEDGEMENTS

This work was supported by the Spanish Ministry of Science and Innovation (MICINN), the ELENA program (CGL2012-31668) and by the FEDER program of the European Union. The cooperation among international partners was supported by the COST Action FP1106, STReESS. We thank three anonymous reviewers and Luka Rejc for improving the English language of this manuscript.

REFERENCES

Alcamo J, Moreno JM, Nováky B, Bindi M, Corobov R, *et al.* 2007. Europe. In: Parry ML, Canziani OF, Palutikof JP, van der Linden PJ, Hanson CE, *et al.* (eds.), Climate Change 2007: Impacts, Adaptation and Vulnerability. Contribution of Working Group II to the Fourth Assessment Report of the Intergovernmental Panel on Climate Change: 541–580. Cambridge University Press, Cambridge, UK.

Baillie MGL & Pilcher JR. 1973. A simple cross-dating program for tree-ring research. Tree-Ring Bull. 33: 7–14.

Barbéro M, Loisel R, Quezel P, Richardson MD & Romane F. 1998. Pines of the Mediterranean basin. In: Richardson DM (ed.), Ecology and biogeography of *Pinus*: 153–170. Cambridge University Press, Cambridge.

Battipaglia G, De Micco V, Brand WA, Linke P, Aronne G, Saurer M & Cherubini P. 2010. Variations of vessel diameter and $\delta^{13}C$ in false rings of *Arbutus unedo* L. reflect different environmental conditions. New Phytol. 188: 1099–1112.

Battipaglia G, Saurer M, Cherubini P, Calfapietra C, McCarthy HR, Norby RJ & Cotrufo MF. 2013. Elevated CO_2 increases tree-level intrinsic water use efficiency: insights from carbon and oxygen isotope analyses in tree rings across three forest FACE sites. New Phytol. 197: 544–554.

Bogino S & Bravo F. 2009. Climate and intra-annual density fluctuations in *Pinus pinaster* subsp. *mesogeensis* in Spanish woodlands. Can. J. For. Res. 39: 1557–1565.

Briffa K, Jones PD, Bartholin TS, Eckstein D, Schweingruber FH, Karlén W, Zetterberg P & Eronen M. 1992. Fennoscandian summers from ad 500: temperature changes on short and long timescales. Clim. Dyna. 7: 111–119.

Campelo F, Gutierrez E, Ribas M, Nabais C & Freitas H. 2007b. Relationship between climate and double rings in *Quercus ilex* from northeast Spain. Can. J. For. Res. 37: 1915–1923.

Campelo F, Nabais C, Freitas H & Gutierrez E. 2007a. Climatic significance of tree-ring width and intra-annual density fluctuations in *Pinus pinea* from a dry Mediterranean area in Portugal. Ann. For. Sci. 64: 229–238.

Campelo F, Vieira J & Nabais C. 2013. Tree-ring growth and intra-annual density fluctuations of *Pinus pinaster* responses to climate: does size matter? Trees 27: 763–772.

Cherubini P, Humbel T, Beeckman H, Gärtner H, Mannes D, *et al.* 2013. Olive tree-ring problematic dating: a comparative analysis on Santorini (Greece). PLoS ONE 8: e54730. doi:10.1371/journal.pone.0054730.

Christensen JH, Hewitson B, Busuioc A, Chen A, Gao X, *et al.* 2007. Regional climate projections. In: Solomon S, Qin D, Manning M, Chen Z, Marquis M, *et al.* (eds.), Climate Change 2007: The physical science basis. Contribution of Working group I to the Fourth Assessment Report of the Intergovernmental Panel on Climate Change: 847–940. Cambridge University Press, Cambridge, UK and New York, USA.

Copenheaver CA, Gärtner H, Shäffer I, Vaccari FP & Cherubini P. 2010. Drought-triggered false ring formation in Mediterranean shrubs. Botany 88: 545–555.

Copenheaver CA, Pokorski EA, Currie JE & Abrams MD. 2006. Causation of false ring formation in *Pinus banksiana*: a comparison of age, canopy class, climate, and growth rate. For. Ecol. Manag. 236: 348–355.

De Luis M, Gričar J, Čufar K & Raventós J. 2007. Seasonal dynamics of wood formation in *Pinus halepensis* from dry and semi-arid ecosystems in Spain. IAWA J. 28: 389–404.

De Luis M, Novak K, Čufar K & Raventós, J. 2009. Size mediated climate-growth relationships in *Pinus halepensis* and *Pinus pinea*. Trees 23: 1065–1073.

De Luis M, Novak K, Raventós J, Gričar J, Prislan P & Čufar K. 2011a. Cambial activity, wood formation and sapling survival of *Pinus halepensis* exposed to different irrigation regimes. For. Ecol. Manag. 262: 1630–1638.

De Luis M, Novak K, Raventós J, Gričar J, Prislan P & Čufar K. 2011b. Climate factors promoting intra-annual density fluctuations in Aleppo pine (*Pinus halepensis*) from semiarid sites. Dendrochronologia 29: 163–169.

De Micco V, Battipaglia G, Brand W, Linke P, Saurer M, Aronne G & Cherubini P. 2012. Discrete versus continuous analysis of anatomical and $\delta^{13}C$ variability in tree rings with intra-annual density fluctuations. Trees 26: 513–524.

De Micco V, Saurer M, Aronne G, Tognetti R & Cherubini P. 2007. Variations of wood anatomy and $\delta^{13}C$ within-tree rings of coastal *Pinus pinaster* showing intra-annual density fluctuations. IAWA J. 28: 61–74.

De Micco V, Zalloni E, Balzano A & Battipaglia G. 2013. Fire influence on *Pinus halepensis*: wood responses close and far from scar. IAWA J. 34: 446–458.

Edmondson JR. 2010. The meteorological significance of false rings in eastern red cedar (*Juniperus virginiana* L.) from the southern great plains, U.S.A. Tree-ring Research 66: 19–33.

Esper J, Cook ER, Krusic PJ, Peters K & Schweingruber FH. 2003. Tests of the RCS method for preserving low-frequency variability in long tree-ring chronologies. Tree-ring Research 59: 81–98.

Fonti P, von Arx G, García-González I, Eilmann B, Sass-Klaassen U, Gärtner H & Eckstein D. 2010. Studying global change through investigation of the plastic responses of xylem anatomy in tree rings. New Phytol. 185: 42–53.

Girard F, Vennetier M, Guibal F, Corona C, Ouarmim S & Herrero A. 2012. *Pinus halepensis* Mill. crown development and fruiting declined with repeated drought in Mediterranean France. Europ. J. For. Res. 131: 919–931.

Harley GL, Grission-Mayer HD, Franklin JA, Anderson C & Kose N. 2012. Cambial activity of *Pinus elliottii* var. *densa* reveals influence of seasonal insolation on growth dynamics in the Florida Keys. Trees 26: 1449–1459.

Herrera S, Gutiérrez JM, Ancel R, Pons MR, Frías MD & Fernández J. 2012. Development and analysis of a 50 year high-resolution daily gridded precipitation dataset over Spain (Spain02). Int. J. Climatol. 32: 74–85.

Hoffer M & Tardif J. 2009. False rings in jack pine and black spruce trees from eastern Manitoba as indicators of dry summers. Can. J. For. Res. 39: 1722–1736.

Holmes R. 1994. Dendrochronology program library user's manual. Laboratory of Tree-Ring Research, University of Arizona, Tucson, USA.

Mamet SD & Kershaw GP. 2013. Age-dependency, climate, and environmental controls of recent tree-growth trends at subarctic and alpine treelines. Dendrochronologia 31: 75–87.

Martin-Benito D, Beeckman H & Canellas I. 2013. Influence of drought on tree rings and tracheid features of *Pinus nigra* and *Pinus sylvestris* in a mesic Mediterranean forest. Europ. J. For. Res. 132: 33–45.

Martinez-Meier A, Sanchez L, Pastorino M. Gallo L & Rozenberg P. 2008. What is hot in tree rings? The wood density of surviving Douglas-firs to the 2003 drought and heat wave. For. Ecol. Manag. 256: 837–843

Nicault A, Alleaume S, Brewer S, Carrer M, Nola P & Guiot J. 2008. Mediterranean drought fluctuation during the last 500 years based on tree-ring data. Clim. Dyna. 31: 227–245.

Novak K, De Luis M, Čufar K & Raventós J. 2011. Frequency and variability of missing tree rings along the stems of *Pinus halepensis* and *Pinus pinea* from a semiarid site in SE Spain. J. Arid Environm. 75: 494–498.

Novak K, De Luis M, Raventos J & Čufar K. 2013. Climatic signals in tree-ring widths and wood structure of *Pinus halepensis* in contrasted environmental conditions. Trees: 27: 927–936.

Olivar J, Bogino S, Spiecker H & Bravo F. 2012. Climate impact on growth dynamics and intra-annual density fluctuations in Aleppo pine (*Pinus halepensis*) tree on different crown classes. Dendrochronologia 30: 35–47.

Panayotov M, Zafirov N & Cherubini P. 2013. Fingerprints of extreme climate events in *Pinus sylvestris* tree rings from Bulgaria. Trees 27: 211–227.

Richardson DM & Rundel PW. 1998. Ecology and biogeography of *Pinus*: an introduction. In: Richardson DM (ed.), Ecology and biogeography of *Pinus*: 3–46. Cambridge University Press, Cambridge, UK.

Rigling A, Waldner PO, Forster T, Bräker O & Pouttu A. 2001. Ecological interpretation of tree-ring width and intraannual density fluctuations in *Pinus sylvestris* on dry sites in the central Alps and Siberia. Can. J. For. Res. 31: 18–31.

Rozas V, García-González I & Zas R. 2011. Climatic control of intra-annual wood density fluctuations of *Pinus pinaster* in NW Spain. Trees 25: 443–453.

Schweingruber FH. 1988. Tree rings, basics and application of dendrochronology. Kluwer Academic Publishers, Dordrecht, Boston, London.

Vieira J, Campelo F & Nabais C. 2009. Age-dependent responses of tree-ring growth and intra-annual density fluctuations of *Pinus pinaster* to Mediterranean climate. Trees 23: 257–265.

Vieira J, Campelo F & Nabais C. 2010. Intra-annual density fluctuations of *Pinus pinaster* are a record of climatic changes in the western Mediterranean region. Can. J. For. Res. 40: 1567–1575.

von Arx G, Kueffer C & Fonti P. 2013. Quantifying plasticity in vessel grouping – added value from the image analysis tool ROXAS. IAWA J. 34: 433–445.

Wegner L, von Arx G, Sass-Klaassen U & Eilmann B. 2013. ROXAS - an efficient and accurate tool to detect vessels in diffuse-porous species. IAWA J. 34: 425–432.

Zubizarreta-Gerendiain A, Gort-Oromi J, Mehtätalo L, Peltola H, Venäläinen A & Pulkkinen P. 2012. Effects of cambial age, clone and climatic factors on ring width and ring density in Norway spruce (*Picea abies*) in southeastern Finland. For. Ecol. Manag. 263: 9–16.

Accepted: 3 September 2013

IAWA Journal 34 (4), 2013: 475–484

BRILL

PLASTIC GROWTH RESPONSE OF
EUROPEAN BEECH PROVENANCES TO DRY SITE CONDITIONS

Srdjan Stojnic[1,*], Ute Sass-Klaassen[2], Sasa Orlovic[1], Bratislav Matovic[1] and Britta Eilmann[2]

[1]University of Novi Sad, Institute of Lowland Forestry and Environment, Antona Cehova 13, 21000 Novi Sad, Republic of Serbia
[2]Wageningen University, Forest Ecology and Management Group, 6700 AA Wageningen, The Netherlands
*Corresponding author: e-mail: srdjan_stojnic@yahoo.com

ABSTRACT

Due to projected global warming, there is a great concern about the ability of European beech to adapt to future climate conditions. Provenance trials provide an excellent basis to assess the potential of various provenances to adjust to given climate conditions. In this study we compared the performance of four European beech (*Fagus sylvatica* L.) provenances growing in a provenance trial at the Fruška Gora Mountain, Serbia. Three of the investigated provenances (Höller-bach and Hasbruch from Germany and Vrani Kamen from Croatia) originate from moist sites, with annual precipitation sums being twice as high as at the provenance trial in Serbia. The performance of these provenances are compared to the growth of the local provenance Fruška Gora which is well adapted to dry site conditions. We analysed tree-ring width, mean vessel area, vessel density and water-conductive area for the period from 2006 to 2012. In spite of differences in climate conditions at their place of origin all beech provenances showed a similar pattern in radial increment. Also the wood- anatomical variables showed similar inter-annual patterns for all provenances and no significant differences between the provenances. This indicates that beech provenances from moist environments can adjust to the relatively dry temperate climate in Serbia.

Keywords: European beech, provenance trial, tree-ring width, wood anatomical structure, phenotypic plasticity.

INTRODUCTION

European beech (*Fagus sylvatica* L.) is one of the major and wide-spread forest tree species in Europe, covering an area of approximately 14 million ha of forest land (von Wühlisch 2010). The expected climatic changes with an increased frequency and duration of intense summer droughts (IPCC 2007) will negatively affect beech, which is known to be sensitive to drought (Rose *et al.* 2009). Thus, the stability and sustain-ability of European beech ecosystems is at risk (von Wühlisch 2004). Therefore it is

© International Association of Wood Anatomists, 2013
Published by Koninklijke Brill NV, Leiden

DOI 10.1163/22941932-00000038

important to know if beech is able to persist future drought events or can even adapt to exacerbating climate conditions.

Existing trees can cope with changing climate conditions through migration, natural selection, or phenotypic plasticity (Nicotra *et al.* 2010). However, species ability to migrate is increasingly limited by fragmentation (Iverson *et al.* 2004). Natural selection on the other hand, might be too slow and ineffective (Rehfeld *et al.* 2002). Thus, phenotypic plasticity could be of primary importance for the performance and survival in a different future climate (Mátyás 2006; Gea-Izquierdo *et al.* 2013).

European beech is known to respond plastically to environmental conditions. Wortemann *et al.* (2011) found no significant difference in cavitation resistance for different beech provenances. Similarly, Meier and Leuschner (2008) found in an experiment that most fine-root traits in beech showed a plastic response when submitted to dry and wet treatments.

In this study we analyse four beech provenances with different origin growing together in a provenance trial at Fruška Gora Mountain. Three provenances originating from mesic sites of Central and Northern Europe (Vrani Kamen from Croatia and Höllerbach and Hasbruch from Germany) were selected and compared with a local provenance (Fruška Gora) on a dry site in Serbia. Tree-ring width, mean vessel area, vessel density and water conductive area were analysed for the period from 2006 to 2012. The study aim was to detect differences between provenances in tree growth and the structure of the water conducting system in response to dry site conditions in a provenance trial at Fruška Gora Mountain.

We hypothesise that beech provenances from moist sites of Central and Northern Europe are more affected by the dry conditions at the trial site in Serbia than the local provenance and will thus show lower productivity rates. This would be in accordance with the results of Eilmann *et al.* (2013) showing that coastal Douglas-fir provenances, originating from moist conditions in the northern part of the distribution range of the species are generally more productive and less drought tolerant than provenances from southern origins. Drought causes a reduction in vessel size (Zhang *et al.* 1992; Sass & Eckstein 1995; Pumijumnong & Park 1999; Arend & Fromm 2007; Eilmann *et al.* 2009). Due to a strong relationship between cell size and its function, even small differences in wood structure between provenances will cause substantial changes in water transport efficiency and safety (Tyree & Zimmermann 2002) and thus in trees' potential to withstand drought. Wider vessels increase water-transportation efficiency (Roderick & Berry 2001; Sperry *et al.* 2006; Anfodillo *et al.* 2013) thus allowing higher stomatal conductance and more carbon gain (Santiago *et al.* 2004). However, large vessels are more vulnerable to drought-induced cavitation (Sperry *et al.* 1994).

A single extreme drought event can cause xylem dysfunction due to cavitation (McDowell *et al.* 2008) substantially reducing plant hydraulic conductance (Tyree & Sperry 1989), with negative consequences on photosynthesis and biomass production (Hölttä *et al.* 2009). Assuming a stronger effect of drought on provenances from mesic sites we expect these provenances to build a less efficient but safer water-conductive system (*i.e.* smaller vessels) leading to lower productivity compared to the local provenance which is well adapted to the generally dry conditions at the trial.

MATERIAL AND METHOD

Study site and plant material

Four provenances of European beech were sampled in a provenance trial close to the village of Stari Ledinci, in the northern part of Serbia. The trial (N 45° 10', E 19° 47') is located at the Fruška Gora Mountain at 370 m a.s.l. on a north-west exposed slope. This provenance trial was established during spring 2007 in the framework of the COST Action E52 by planting 3-year-old saplings of 25 European beech provenances. The trial is arranged in a randomised complete block design. Fifty saplings were planted per plot with 1×2 m spacing. Around the trial, a buffer strip of 2 rows with the local provenance (Fruška Gora) was planted to avoid edge effects (Speer 2010). The soil is an acid brown soil with a pH of 5.4.

The climate is temperate continental with a mean annual temperature of 11.1 °C and annual precipitation sum of 624 mm (Fig. 1). Mean air temperature during the vegetation period (April–September) is 17.8 °C, while the sum of precipitations for the same period amounts to 369 mm. The climate records are from the weather station Rimski Sancevi (N 45° 20', E 19° 51'; 84 m a.s.l.), at 30 km distance from the trial (http://www.hidmet.gov.rs). Temperature and precipitation have been averaged for the time period between 1966 and 2004.

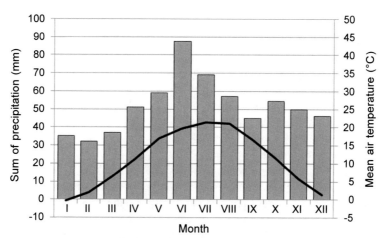

Figure 1. Climate diagram according to Walter and Lieth (1967) for the weather station Rimski Sancevi (N 45° 20', E 19° 51'; 84 m a.s.l.) (norm period: 1966–2004).

To describe water availability since establishment of the site the De Martonne aridity index (Am) (De Martonne 1926) was calculated from annual mean temperature T [°C] and annual precipitation P [cm] from the weather station Rimski Sancevi for the period from 2007 to 2012.

$$Am = \frac{P}{T + 10}$$

Low values of the Am aridity index refer to low water availability while well watered conditions are expressed by high values.

Table 1. General data about European beech provenances involved in study.

Provenance	Country of origin	Longitude	Latitude	Altitude (m)	Annual mean air temp. (°C)	Annual sum of precip. (mm)
Vrani Kamen	Croatia	17° 19'	45° 37'	600	8.5	972
Fruška Gora	Serbia	19° 55'	45° 10'	370	11.1	624
Höllerbach	Germany	13° 14'	49° 1'	755	5.0	1200
Hasbruch	Germany	8° 26'	53° 8'	35	8.6	760

Four provenances originating from Croatia (Vrani Kamen), Germany (Höllerbach and Hasbruch) and Serbia (Fruška Gora), were selected, covering a gradient in climate conditions from the north to the south of Europe (Table 1). Hasbruch originates from the northern part of Germany and it is the most northern provenance represented in the trial, coming from a very low altitudinal site (35 m a.s.l.). Mean air temperature at the site is 8.6 °C, while the annual sum of precipitation amounts to 760 mm. Another German provenance, Höllerbach, originates from the southern part of Germany. This provenance originates from the highest altitude, *i.e.* 755 m a.s.l., with annual sum of precipitation as high as 1200 mm and a relatively low mean annual air temperature of 5.0 °C. The provenance Vrani Kamen originates from the north-eastern part of Croatia and could be described as Central-European provenance related to the geographical position of Croatia. At this location the mean annual temperature is 8.5 °C and the annual sum of precipitation is 972 mm. The local provenance Fruška Gora originates from the northern part of Serbia. Site conditions at the locality can be described as xeric for beech with a mean annual air temperature at this mountain site (370 m a.s.l.) getting to on average 11.1 °C, while the annual sum of precipitation is only 624 mm.

Sampling and data analysis

Five dominant trees per provenance were sampled by taking increment cores close to the stem base but high enough to escape influence of the root collar. Due to the small size of the trees and in order to reduce the damage on the trees due to sampling, only one increment core went through the entire tree. Unfortunately the second half of the core had to be neglected during tree-ring and wood-anatomical analysis due to low wood quality of the piece as a result of a lot of cracks and breakages.

In order to obtain smooth wood surfaces for tree-ring analysis, the cores were planed with a core-microtome (WSL, Birmensdorf, Switzerland). For a better contrast, the cell lumina were filled with white chalk. Analysis of tree-ring width was performed using a combination of a Lintab digital positioning table and the software TSAP-Win (both Rinntech, Germany). Single tree-ring series were cross-dated visually and averaged into provenance chronologies.

Thin sections of all cores with a thickness of 10 μm were cut using a field sliding microtome (WSL, Switzerland). To increase the contrast between the cell walls and the cell lumina thin sections were stained using an alloy of safranin-O and astra-blue (150 mg astrablue, 40 mg safranin and 2 ml acetic acid in 100 ml distilled water). To produce permanent samples the thin sections were dehydrated in an alcohol concentration gradient (50%, 95% and absolute ethanol), then in Roticlear® and embedded in Roti®-Mount (both Carl Roth, Germany). Pictures of the thin sections were taken with a Leica DFC320 camera associated to an optical microscope (Leica DM2500) and the software Leica Application Suite v3.8 (all products Leica, Germany) with a 2.5× objective. Using the software Autostitch all photographs taken per core were merged to one big picture. Analysis of mean vessel area (μm^2) and vessel density were performed with the free web software ImageJ (http://rsb.info.nih.gov/ij/).

Percentage of water conducting area was calculated based on values of total vessel area and tree-ring area according to formula:

$$\text{Percentage of water} - \text{conducting area} = \frac{\text{Total vessel area}}{\text{Tree ring area}} \times 100 \ (\%)$$

Statistical analyses were carried out using software Statistica 12 (StatSoft, Inc.). To test for differences between provenances regarding tree-ring width and wood-anatomical variables an ANOVA followed by a Scheffe's post-hoc test was performed. Linear-regression analysis was performed between wood-anatomical variables and tree-ring width, using the mean values of each variable.

RESULTS

All beech provenances show a similar pattern in radial increment with a distinct plant shock (low radial increment) in the first years after planting and marginal differences in the level of increment (Fig. 2). Values of Am reveal that climate during the observed period varied in a broad range (*i.e.* 2011 and 2012 were "semi dry" years, while the 2007 and 2010 were "humid" years).

Even though the southern provenances (Vrani Kamen and Fruška Gora) tend to have a higher tree-ring width (TRW), compared to the provenances from the North, these differences were not statistically significant. Provenance Hasbruch showed the lowest increment of all provenances, but still the differences between the provenances were not significant according to the ANOVA.

Mean vessel area (MVA) and vessel density (VD) showed similar development for all provenances within the observation period. Like for TRW, district response to the year of trial establishment (plant shock) in 2007 became apparent resulting in increasing VD and decreasing MVA. The lowest MVA was found in Hasbruch, the highest in Vrani Kamen and Fruška Gora (Fig. 2) but also here the differences between provenances were not significant. Regarding VD, only Vrani Kamen showed slightly higher values than the other provenances (Fig. 2). Vrani Kamen showed the highest conductive area (WCA) of all provenances while Hasbruch and Fruška Gora showed the lowest values.

A detailed climate-growth analysis revealing differences between the provenances in climate responses was not possible due to the young age of the plants and superimposing effects, like the plant shock.

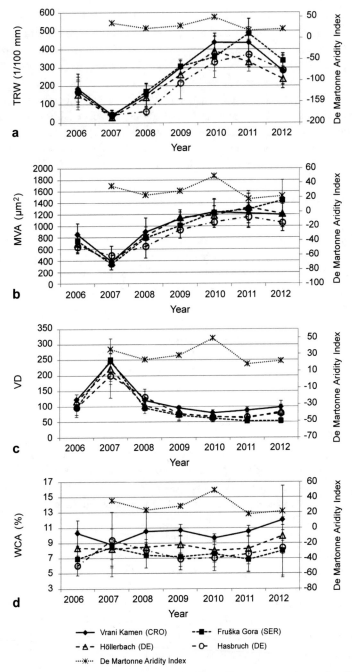

Figure 2. Variation in: **a**: tree-ring width [1/100 mm], **b**: mean vessel area [μm²], **c**: vessel density, and **d**: water conductive area [%], separately for the four provenances. De Martonne Aridity Index (Am) was calculated for the time period from 2007 to 2012. Error bars give the standard deviation.

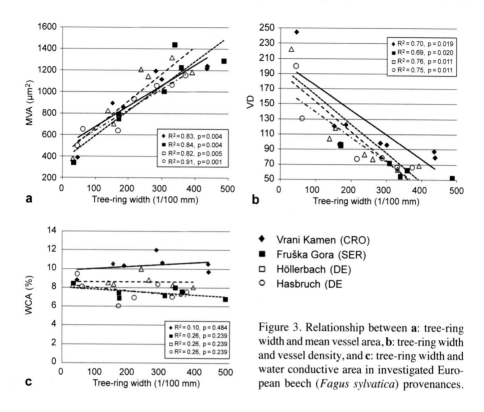

◆ Vrani Kamen (CRO)
■ Fruška Gora (SER)
□ Höllerbach (DE)
○ Hasbruch (DE

Figure 3. Relationship between **a**: tree-ring width and mean vessel area, **b**: tree-ring width and vessel density, and **c**: tree-ring width and water conductive area in investigated European beech (*Fagus sylvatica*) provenances.

Wood-anatomical variables were strongly related to tree-ring width in all provenances. MVA clearly increased with increasing tree-ring width (p < 0.01; Fig. 3) while vessel density was negatively related to tree-ring width and VD was reduced in wider tree rings (p < 0.5; Fig. 3). Only the relationship between tree-ring width and WCA was not significant.

DISCUSSION

All beech provenances responded quite similarly to the site conditions prevailing in the Serbian trial. Even Höllerbach, the provenance originating from a region with approximately twice as much precipitation per year compared to the conditions in the trial, showed similar levels of increment and a similar wood structure like the local provenance. This points to a high potential of beech trees to adjust their wood-anatomical variables to local pedo-climatic conditions, allowing trees from moist sites to form an equally efficient water-conducting system as the local beech provenances resulting in similar levels of increment. Thus even provenances from moist sites have the ability to adjust well to the relatively dry continental conditions in Serbia. Plastic response of beech to drought has already been documented for root traits (Meier & Leuschner 2008) and biomass partitioning (Löf *et al.* 2005).

The ability of beech provenances from moist sites to respond similarly to the dry climate conditions as the local provenance should be taken into the account when modelling the future distribution of beech under climate change. Recent predictions expect beech to lose habitats in the south and south-eastern edge of the present distribution, due to changing climate conditions (Kramer *et al.* 2010). Still the same authors stated that the adaptive responses of key functional traits should not be ignored when estimating the potential of beech to withstand exacerbating climate conditions in the future. Substantial phenotypic plasticity might allow species to survive even at marginal sites, by expressing a range of different phenotypes depending on environmental conditions (Sultan 2000; Benito Garzón *et al.* 2011). Thus, the potential of European beech to adapt to drought might be underestimated.

Even though our study showed that all beech provenances respond plastically when planted to a dry site, this does not necessarily mean that existing beech stands can easily adjust to sudden changes in environmental conditions since the tree architecture of the trees is not adjusted to changing climate conditions (Eilmann *et al.* 2009; Eilmann & Rigling 2012). This indicates the importance of additional analysis concerning *in situ* adaptation to future conditions in natural beech populations.

Our study revealed a positive significant relationship between tree-ring width and MVA and a negative one with VD (Fig. 3). These trends are in accordance with the results of Pourtahmasi *et al.* (2011) on *Fagus orientalis*. The positive relationship between vessel area and tree growth can be explained by the positive effect of vessel area on transport efficiency, allowing a higher stomatal conductance and thus higher photosynthetic activity and resulting in increased productivity rate (Poorter *et al.* 2010).

In conclusion, our results showed that provenances originating from mesic sites (Vrani Kamen, Hasbruch and Höllerbach), thanks to high plastic response, can adjust to the generally dry conditions in Serbia and showed similar productivity rates and wood structures like the local provenance (Fruška Gora). It would be desirable to extend this research towards the actual response of beech provenances to years where prolonged periods of drought prevailed and study effects on ring width and wood structure and the underlying physiological processes in more detail.

ACKNOWLEDGEMENTS

This study was realised as part of the short-term scientific mission: "Detecting drought traces in the wood structure of contrasting European beech provenances" (COST STSM Reference number: COST-STSM-FP1106-12371) at the DendroLab, Wageningen University. The study was financed by the COST Action FP1106: STReESS - Studying Tree Responses to extreme Events: a SynthesiS.

REFERENCES

Anfodillo T, Petit G & Crivellaro A. 2013. Axial conduit widening in woody species: a still neglected anatomical pattern. IAWA J. 34: 352–364.

Benito Garzón M, Alía R, Robson M & Zavala MA. 2011. Intra-specific variability and plasticity influence potential tree species distributions under climate change. Global Ecol. Biogeogr. 20: 766–778.

De Martonne E. 1926. Une nouvelle function climatologique: l'indice d'aridité. Meteorologie 2: 449–458.

Eilmann B, de Vries SMG, den Ouden J, Mohren GMJ, Sauren P & Sass UGW. 2013. Origin matters! Difference in drought tolerance and productivity of coastal Douglas-fir (*Pseudotsuga menziesii* (Mirb.)) provenances. Forest Ecol. Manag. 302: 133–143.

Eilmann, B & Rigling A. 2012. Tree-growth analyses to estimate tree species' drought tolerance. Tree Physiol. 3: 178–187.

Eilmann B, Zweifel R, Buchmann N, Fonti P & Rigling A. 2009. Drought-induced adaptation of the xylem in Scots pine and pubescent oak. Tree Physiol. 29: 1011–1020.

Fritts HC. 2001. Tree rings and climate. Academic Press, London.

Gea-Izquierdo G, Battipaglia G, Gärtner H & Cherubini P. 2013. Xylem adjustment in *Erica arborea* to temperature and moisture availability in contrasting climates. IAWA J. 34: 109–126.

Hölttä T, Cochard H, Nikinmaa E & Mencuccini M. 2009. Capacitive effect of cavitation in xylem conduits: results from a dynamic model. Plant Cell Environm. 32: 10–21.

IPCC. 2007. Summary for policymakers. In: Salomon S, Qin D, Manning M, Chen Z, Marquis M, Averyt KB, Tignor M & Miller HL (eds.), Climate change 2007: The physical science basis. Contribution of working group I to the fourth assessment report of the intergovernmental panel on climate change. Cambridge Univ. Press, Cambridge, UK and New York, NY, USA.

Iverson LR, Schwartz MW & Prasad AM. 2004. How fast and far might tree species migrate in the eastern United States due to climate change? Global Ecol. Biogeogr. 13: 209–219.

Kramer K, Degen B, Buschbom J, Hickler T, Thuiller W, Sykes MT & de Winter W. 2010. Modelling exploration of the future of European beech (*Fagus sylvatica* L.) under climate change - range, abundance, genetic diversity and adaptive response. Forest Ecol. Manag. 259: 2213–2222.

Löf M, Bolte A & Welander NT. 2005. Interacting effects of irradiance and water stress on dry weight and biomass partitioning in *Fagus sylvatica* seedlings. Scand. J. Forest Res. 20: 322–328.

Mátyás C. 2006. Migratory, genetic and phenetic response potential of forest tree populations facing climate change. Acta Silv. Lign. Hung. 2: 33–46.

McDowell N, Pockman WT, Allen CD, Breshears DD, Cobb N, Kolb T, Plaut J, Sperry J, West A, Williams DG & Yepez EA. 2008. Mechanisms of plant survival and mortality during drought: why do some plants survive while others succumb to drought? New Phytol. 178: 719–739.

Meier IC & Leuschner C. 2008. Genotypic variation and phenotypic plasticity in the drought response of fine roots of European beech. Tree Physiol. 28: 297–309.

Nicotra AB, Atkin OK, Bonser SP, Davidson AM, Finnegan EJ, Mathesius U, Poot P, Puruganan MD, Richards CL, Valladares F & van Kleunen M. 2010. Plant phenotypic plasticity in a changing climate. Trends Plant Sci. 15: 684–692.

Poorter L, McDonald I, Alarcon A, Fichtler E, Licona JC, Pena-Claros M, Sterck F, Villegas Z & Sass-Klaassen U. 2010. The importance of wood traits and hydraulic conductance for the performance and life history strategies of 42 rainforest tree species. New Phytol. 185: 481–492.

Pourtahmasi K, Lotfiomran N, Bräuning A & Parsapajouh D. 2011. Tree-ring width and vessel characteristics of Oriental beech (*Fagus orientalis*) along an altitudinal gradient in the Caspian forests, northern Iran. IAWA J. 32: 461–473.

Pumijumnong N & Park WK. 1999. Vessel chronologies from teak in northern Thailand and their climatic signal. IAWA J. 20: 285–294.

Rehfeld GE, Tchebakova NM, Parfenova YI, Wykoff WR, Kuzmina NA & Milyutin LI. 2002. Intraspecific responses to climate in *Pinus sylvestris*. Glob. Chang. Biol. 8: 912–929.

Roderick ML & Berry SL. 2001. Linking wood density with tree growth and environment: a theoretical analysis based on the motion of water. New Phytol. 149: 473–485.

Rose L, Leuchner C, Köckemann B & Buschmann H. 2009. Are marginal beech (*Fagus sylvatica* L.) provenances a source for drought tolerant ecotypes? Eur. J. For. Res. 128: 335–343.

Sass U & Eckstein D. 1995. The variability of vessel size in beech (*Fagus sylvatica* L.) and its ecophysiological interpretation. Trees - Struct. Funct. 9: 247–252.

Speer JH. 2010. Fundamentals of tree-ring research. The University of Arizona Press, Tuscon.

Sperry JS, Nichols KL, Sullivan JE & Eastlack SE. 1994. Xylem embolism in ring-porous, diffuse-porous and coniferous trees in northern Utah and interior Alaska. Ecology 75: 1736–1752.

Sultan SE. 2000. Phenotypic plasticity for plant development, function and life history.Trends Plant Sci. 5: 537–542.

Tyree MT & Sperry JS. 1989. Vulnerability of xylem to cavitation and embolism. Annu. Rev. Plant Phys. 40: 19–38.

Tyree MT & Zimmermann MH. 2002. Xylem structure and the ascent of sap. Springer Verlag, Berlin.

von Wühlisch G. 2004. Series of International Provenance Trials of European Beech. In: Sagheb-Talebi K, Madsen P & Terazawa K (eds.), Proceedings from the 7th International Beech Symposium IUFRO Research Group 1.10.00: Improvement and Silviculture of Beech: 135–144. Research Institute of Forests and Rangelands (RIFR), Iran.

von Wühlisch G. 2010. Introductory note. In: Frýdl J, Novotný P, Fennessy J & von Wühlisch G (eds.), Genetic resources of beech in Europe – current state. Communicationes Instituti Forestalis Bohemicae 25: 8–9. Forestry and Game Management Research Institute, Czech Republic.

Walter H & Lieth H. 1967. Climate diagram world atlas. G. Fischer, Jena.

Wortemann R, Herbette S, Barigah TS, Fumanal B, Alia R, Ducousso A, Gomory D, Roeckel-Drevet P & Cochard H. 2011. Genotypic variability and phenotypic plasticity of cavitation resistance in *Fagus sylvatica* L. across Europe. Tree Physiol. 31: 1175–1182.

Accepted: 6 September 2013

IAWA Journal 34 (4), 2013: 485–497

EVALUATING THE WOOD ANATOMICAL AND DENDROECOLOGICAL POTENTIAL OF ARCTIC DWARF SHRUB COMMUNITIES

Fritz Hans Schweingruber[1], Lena Hellmann[1,2], Willy Tegel[3], Sarah Braun[3], Daniel Nievergelt[1] and Ulf Büntgen[1,2,4,*]

[1]Swiss Federal Research Institute, WSL, Zürcherstraße 111, CH-8903 Birmensdorf, Switzerland
[2]Oeschger Centre for Climate Change Research, Zähringerstraße 25, CH-3012 Bern, Switzerland
[3]Institute for Forest Growth IWW, University of Freiburg, Tennebacherstraße 4, D-79106 Freiburg, Germany
[4]Global Change Research Centre AS CR, v.v.i., Bělidla 986/4a, CZ-60300 Brno, Czech Republic
*Corresponding author; e-mail: buentgen@wsl.ch

ABSTRACT

Supplementing broader-scale dendroecological approaches with high-resolution wood anatomical analyses constitutes a useful technique to assess spatiotemporal patterns of climate-induced growth responses in circumpolar tundra vegetation. A systematic evaluation of dendrochronological and wood anatomical features in arctic dwarf shrubs is, however, still missing. Here, we report on nearly thousand samples from ten major dwarf shrub species that were collected at 30 plot-sites around 70° N and 22° W in coastal East Greenland. Morphological root and stem characteristics, together with intra-annual anatomical variations are outlined and the potential and limitation of ring counting is stressed. This study further demonstrates the possibility to gain annually resolved insight on past dry matter production and carbon allocation in arctic (and alpine) environments well beyond northern (and upper) treelines, where vegetation growth is particularly sensitive to environmental change.

Keywords: Anatomical features, arctic and alpine ecotones, climate change, dendroecology, Greenland, life forms, tundra vegetation, wood anatomy.

INTRODUCTION

Arctic and alpine environments are among the most sensitive regions regarding their reactions to climate change (*e.g.* Serreze & Francis 2006; Kaufman *et al.* 2009; Pauli *et al.* 2012). Increasing temperatures already affected vegetation cover and species composition (*e.g.* Sturm *et al.* 2001; Verbyla 2008; Macias-Fauria *et al.* 2012), and also left pronounced fingerprints in the annual growth rings of trees and shrubs (*e.g.* Büntgen & Schweingruber 2010). The vast majority of high-northern and higher elevation ecotones are characterized by harsh climatological and pedological conditions, which restrict radial stem thickening to fairly short vegetation periods. At the same time dwarf shrubs are well adapted to survive under such environmental extremes, and

© International Association of Wood Anatomists, 2013
Published by Koninklijke Brill NV, Leiden

DOI 10.1163/22941932-00000039

knowledge on their growth behaviour describes a pending challenge to account for possible effects associated with predicted climate change, *e.g.* tundra expansion and carbon sequestration (McGuire *et al.* 2010; Elmendorf *et al.* 2012; Pearson *et al.* 2013).

Early pioneering studies focused on a better understanding of how low arctic temperatures and short vegetation periods decrease radial growth of Greenlandic dwarf shrubs, and at the same time increase plant longevity (Kraus 1873). Comparison of different dwarf shrubs from high- and low-elevation settings in Switzerland, Germany, Russia and the Himalaya confirmed these results (Kihlmann 1890; Rosenthal 1904; Kanngiesser 1914). Miller (1975) published detailed wood anatomical accounts of Greenlandic dwarf shrubs.

Dendroecological research on dwarf shrubs, however, almost stagnated during most of the last century, despite a few exceptions (*e.g.* Good 1927; Molisch 1938; Parsons *et al.* 1994), and a rediscovery during the past decade. An emerging community now concentrates on revealing climatological and ecological information preserved in annual rings, not only based on field studies but also using experimental setups (*e.g.* Schweingruber & Poschlod 2005; Schmidt *et al.* 2006; Bär *et al.* 2006, 2008; Bär & Löffler 2007; Hallinger *et al.* 2010; Blok *et al.* 2011; Myers-Smith *et al.* 2011; Buizer *et al.* 2012; Weijers *et al.* 2012). See also http://shrubhub.biology.ualberta.ca/ for a detailed literature overview.

Technical and economical limitations related to collecting, preparing and analyzing a sufficiently large number of samples, however, hampered the creation of well replicated and long enough wood anatomical and dendrochronological dwarf shrub records. At the same time this drawback stimulated a vital discussion about possible crossdating trials associated with imprecise ring boundaries and irregular growth disturbances (*e.g.* Büntgen & Schweingruber 2010; Hallinger & Wilmking 2011; Wilmking *et al.* 2012; Buchwal *et al.* 2013). Recent research endeavours at the interface of wood anatomy and dendroecology/climatology underscore the urgent need to enhance our understanding of wood anatomical structures to facilitate the distinction of annual rings.

Here, we introduce a suite of anatomical features observed in nearly thousand specimens from ten dwarf shrub species in coastal East Greenland. The discussion is placing our anatomical results in an ecological context, with emphasis on the potential of Arctic dwarf shrubs to allow past changes in their environment to be reconstructed. We ultimately suggest that species-specific anatomical characteristics in relation to site-specific conditions should be considered prior to the measurement and subsequent crossdating of ring widths.

MATERIAL AND METHODS

We sampled 945 dwarf shrub individuals from ten dominant species (*Arctostaphylos alpina, Cassiope hynoides, Cassiope tetragona, Empetrum nigrum, Dryas octopetala, Rhododendron lapponicum, Vaccinium uliginosum, Betula nana, Salix arctica,* and *Salix herbacea*) at 30 plot-sites at the eastern coastline of Greenland near Scoresbysund (*i.e.* Ittoqqortoormiit, 70° 26' 6" N, 21° 58' 100" W). The natural distribution of these species is circumpolar, with some genera also occurring in alpine environments across Eurasia and North America (Hultén 1968). All sampling plots, located between 5 and 320 m asl on permafrost, are either characterized by crystalline orthogneiss with large granite boulders (22 plots) or sedimentary sandstone (8 plots) that creates relatively flat surfaces with overall lower nutrient levels. The short vegetation period is constrained by a cold and dry climate. Mean summer temperature and total precipitation is 1.9 °C and 94 mm, respectively (calculated for June–August over the period 1950–1980 and using monthly resolved GHCN Version 3 data) (Lawrimore *et al.* 2011).

We performed a rigorous plot-site sampling of the complete above- and belowground stem section of each existing specimen within a radius of ~20 m, to capture the full variety of prevailing plant sizes and ages. This strategy ensures obtaining the thickest part of highly branched plants and root systems, often being hidden in the ground or beneath rocks. Inclusion of the taproot that contains the maximum number of rings within a plant is essential for all further analyses, as it enables ring properties and plant ages to be most accurately determined.

After labeling and archiving this unique collection, several cross sections per sample were prepared using sliding microtomes with disposable blades. Unstained high-quality sections, solely preserved in glycerol, were already suitable for ring counting. Double staining with safranin and astrablue, however, further visualized a variety of anatomical cell structures (Schweingruber *et al.* 2008; Gärtner & Schweingruber 2013).

Additional sample preservation with Nawashin-solution and extra staining with Picric-anilinblue allowed the visualization of cell contents as well as cell nuclei, and thus the estimation of cell longevity. After being covered with Nawashin for ten minutes, the sections must be washed with water before simultaneously being stained with safranin-astrablue, as well as before ultimately being stained with Picric-anilinblue. Short heating up to 80 °C prepares slides for their usual dehydration and embedding process with ethanol, xylene and 'Canada' balsam (Schweingruber *et al.* 2008). Preserving and dye solutions are composed as follows: Nawashin-solution consists of ten parts of 1% chromic acid, four parts 4% formaldehyde and one part acetic acid (Purvis *et al.* 1964). Safranin-dye entails 0.8 g of safranin powder in 100 ml of distilled water. Astrablue-dye contains 0.5 g of astrablue powder in 100 ml distilled water, and 2 ml acidic acid. Picric-anilinblue-dye is created from one part saturated anilinblue and four parts saturated Picric-acid (trinitrophenol), dissolved in 95% ethanol. The above processing steps, all performed at WSL, finally resulted in 871 samples for which ring counts and subsequent longevity estimates were performed (see Table 1 for a detailed overview).

For each species, we selected microscopic photographs (Fig. 1–5) showing a ring sequence of optimal growth to highlight its dendrochronological capacity (Fig. 1–5a,d), a section of reduced growth to address possible counting and measuring problems (Fig. 1–5b, e), and a peripheral stem with bark to illustrate typical bark features (Fig. 1–5c, f). These high-resolution images help to display the rich anatomical range among the different samples. Species-specific bark thickness and characteristics were additionally recorded to reveal information on their ecological function.

Table 1. Species-specific dwarf shrub characteristics.

Leaf persistence: d = deciduous, e = evergreen.

Ring distinctness: a = very distinct, number of rings corresponds with the true age, good cross-
    dating; b = uncertain determination of plant tissue, total ring number only possible on whole
    cross sections, weak crossdating; c = uncertain age determination, number of counted rings
    only estimated, crossdating impossible; d = ring counting impossible.

Average ring width: values correspond to 945 recorded individuals from the high Arctic, values
    might be different for other sites but relations between species remain.

Number of counted rings: values refer to minimum/mean/maximum estimates.

Porosity: d = diffuse porous, sr = semi-ring porous.

Species	Number of samples	Leaf persistence	Ring distinctness (lsrge rings)	Ring distinctness (small rings)	Average ring width (mm)	Number of counted rings	Porosity
Arctostaphylos alpina	79	d	a & b	c	< 0.20	20/49/114	sr
Betula nana	90	d	a	c	< 0.10	18/62/165	d
Cassiope hypnoides	12	d	b	c	< 0.20	33/54/75	d
Cassiope tetragona	146	e	c	d	< 0.05	13/54/126	d
Dryas octopetala	93	d	a & b	c	< 0.10	33/85/154	d
Empetrum nigrum	32	e	a & b	c	< 0.20	30/62/101	sr
Rhododendron lapponicum	117	d	b	d	< 0.20	1/71/204	d
Salix arctica	156	d	b	c	< 0.10	17/64/197	sr
Salix herbacea	43	d	c	d	< 0.20	4/13/23	sr
Vaccinium uliginosum	103	d	c	d	< 0.20	14/51/96	d

RESULTS AND DISCUSSION

Distinct annual ring boundaries are found in all ten arctic dwarf shrub species (Fig.
1–5). The overall suitability of ring counting, measuring and crossdating, however,
varies significantly between individual plants of the same species. A selection of most
relevant anatomical characteristics within our sample collection is summarized in
Table 1. Information on specific criteria of leaf persistence of deciduous and evergreen
species, the distinctiveness of small and large rings (</>0.5 mm), the average ring
width of all plants per species (mm), and the porosity of growth rings (diffuse or
semi-ring porous) are summarized. This table contains relevant insight to evaluate the
dendrochronological potential of the ten species, which must be understood as a combination of the mean and maximum number of counted rings per species, as well as the
distinctiveness of their ring boundaries. While the first criterion implies the potential
to develop long chronologies, the second parameter infers a benchmark for crossdating trials. Anatomical details for tangential and radial sections are also provided in
Greguss (1945), Miller (1975), Schweingruber (1995), and Benkova & Schweingruber
(2004).

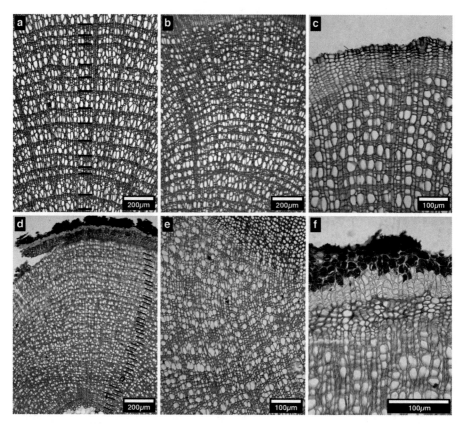

Figure 1. a–c: *Arctostaphylos alpina* – d–f: *Cassiope tetragona*.

Average ring width ranges from < 0.05 mm to 1.0 mm. Ring distinctness is determined by the ring size. Small rings are never clearly distinguishable, often leading to rather rough age estimates. Very limited growth conditions often lead to varying numbers of rings at different radii on cross sections, especially in the aboveground stem sections. Most accurate ring numbers can be achieved by serial sectioning from the dominant thickest root to the tip of the shoot (Büntgen & Schweingruber 2010; Buchwal *et al.* 2013). Although crossdating would yield the correct plant age, it was not realized in this study since we did not aim to ultimately date our material. It is therefore important to note that all age information given here must always be considered as an estimated minimum number of the preserved tissue part at a specific plant section.

Characteristic wood anatomical features allow the assignment to different plant organs: stems and shoots are defined by the existence of pith, whereas roots typically do not contain pith (Fig. 6a, b). Adventitious shoots are not distinguishable from primary shoots and hence hamper any exact age determination of the plant as a whole (de Witte & Stöcklin 2010).

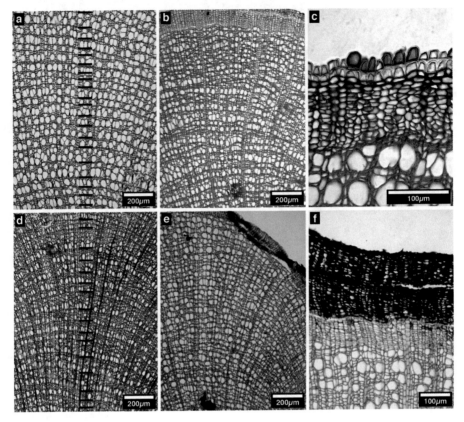

Figure 2. a–c: *Empetrum nigrum* – d–f: *Dryas octopetala*.

Extreme site conditions cause prostrate stems with reduced growth at the upper and enhanced growth at the lower side, resulting in eccentric pith positions. Different degrees of pith eccentricity were classified: centric (Fig. 6c), slightly eccentric (Fig. 6d), extremely eccentric (Fig. 6e), and fully eroded pith (Fig. 6f). Missing or wedging rings are less likely at the longest radius, making corresponding ring counting and ring-width measurements most reliable. However, totally eroded pith might still impede the results and missing years due to wedging rings can never be excluded, even at the longest radius. Centric and slightly eccentric pith positions are typical for *Cassiope tetragona*, *Vaccinium* sp., *Rhododendron lapponicum*, *Betula nana*, and *Salix herbacea*, while extremely eccentric and eroded pith positions are characteristic for *Dryas octopetala*. Only *Betula nana* and *Salix* sp. have tension wood as additional means to compensate mechanical instability.

Sampling during the growing season results in incompletely formed and lignified outermost rings (Fig. 7a, b). Therefore beginning and end of plant growth at different sites within a region can be estimated by comparing the outermost rings of individuals.

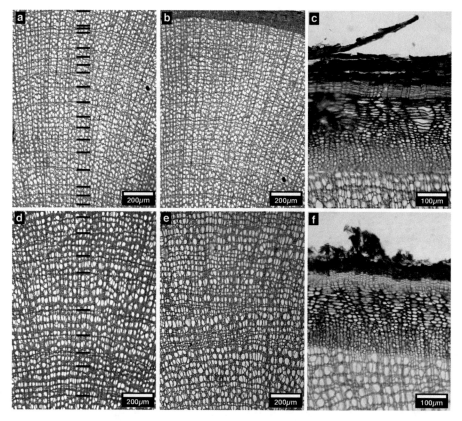

Figure 3. a–c: *Rhododendron lapponicum* – d–f: *Vaccinium uliginosum*.

The amount of living parenchyma (cells with protoplasts) (Fig. 7c) indicates the reactivity of secondary xylem to wounding or biological attack and its ability to store and mobilize metabolites. Extremely long-living parenchyma cells are characteristic for all stems herein analyzed (up to ~204 years for *Rhododendron lapponicum*). Living cells can store carbohydrates and react to mechanical disturbances by callus formation (Fig. 7d) or to fungal attack by barrier zone building (Fig. 7e), while cells in dead parts contain phenols or tyloses. *Salix sp.* is the only species generating typical dark-colored heartwood. All other species show different chemical defense systems. Defense zones against fungal decay are represented by the irregular occurrence of tyloses and phenols (Fig. 7f).

In light of the above, it becomes obvious that the most efficient method for dendroecological studies of dwarf shrubs is annual ring counting and ring-width measuring based on micro sections. Sliding microtomes enable the preparation of high-quality cross sections in reasonable time, and therefore facilitate the production of long and well-replicated datasets. New staining techniques with different dyes allow straightforward analyses of diverse anatomical features on thin sections, including longevity of nuclei

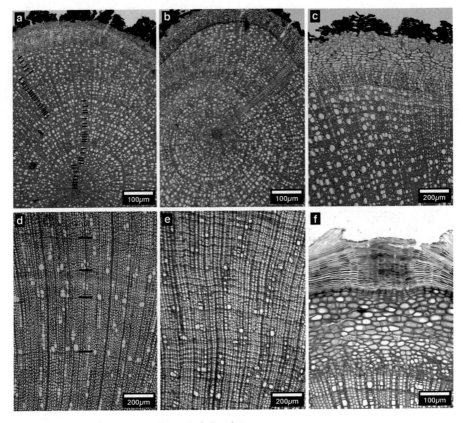

Figure 4. a–c: *Vaccinium myrtillus* – d–f: *Betula nana*.

in cells. The amount of work is reduced while the outcome is increased. Although this study focused on a selection of arctic dwarf shrubs only, the main anatomical features observed are likely also representative for other species growing in alpine environments.

ACKNOWLEDGEMENTS

This study is part of the ongoing 'Arctic driftwood' project supported by the Eva Mayr-Stihl Foundation. U.B. also obtained financial support from the Czech project 'Building up a multidisciplinary scientific team focused on drought' (No. CZ.1.07/2.3.00/20.0248). Albena Ivanova and Loic Schneider contributed lab work.

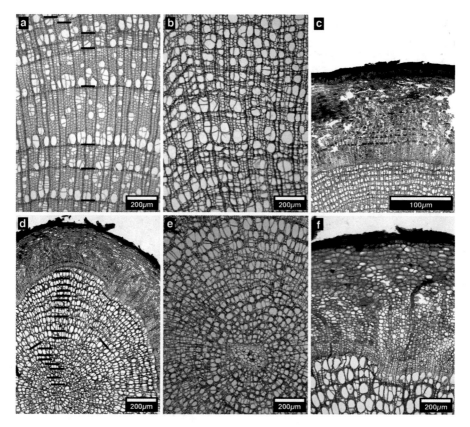

Figure 5. a–c: *Salix arctica* – d–f: *Salix herbacea*.

REFERENCES

Bär A, Bräuning A & Löffler J. 2006. Dendroecology of dwarf shrubs in the high mountains of Norway – A methodological approach. Dendrochronologia 24: 17–27.

Bär A & Löffler J. 2007. Ecological process indicators used for nature protection scenarios in agricultural landscapes of SW Norway. Ecological Indicators 7: 396–411.

Bär A, Pape R, Bräuning A & Löffler J. 2008. Growth-ring variations of dwarf shrubs reflect regional climate signals in alpine environments rather than topoclimatic differences. J. Biogeogr. 35: 625–636.

Benkova VE & Schweingruber FH. 2004. Anatomy of Russian woods. An atlas for the identification of trees, shrubs and dwarf shrubs and woody lianas from Russia. Haupt, Bern.

Blok D, Sass-Klaassen U, Schaepman-Strub G, Heijmans MMPD, Sauren P & Berendse F. 2011. What are the main climate drivers for shrub growth in Northeastern Siberian tundra? Biogeosciences 8: 771–799.

Buchwal A, Rachlewicz G, Fonti P, Cherubini P & Gärtner H. 2013. Temperature modulates intra-plant growth of *Salix polaris* from a high Arctic site (Svalbard). Polar Biology. 06/2013; DOI 10.1007/s00300-013-1349-x.

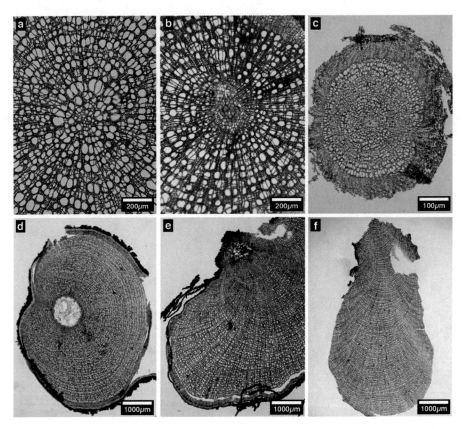

Figure 6. a: Root of *Salix herbacea*, radial growth in the center of the root starts with fibers and vessels. – b: Shoot of *Salix herbacea*, radial growth in the center of the root starts around a non-lignified parenchymatous pith and primary xylem, which is a remnant of primary growth. – c: Centric stem of *Salix herbacea*. – d: Eccentric stem of *Cassiope tetragona*, approximately 35 rings occur on the longest and only 12 rings on the shortest radius, at least 20 rings are wedging out. – e: Extremely eccentric stem of *Dryas octopetala*, at the bottom side occur approximately 100 and at the upper side not more than 5 rings. – f: Extremely eccentric stem of *Dryas octo-petala*, growth occurs only at the bottom side, all rings are laterally wedging out, with the pith being eroded. – a–f are stained with safranin-astrablue.

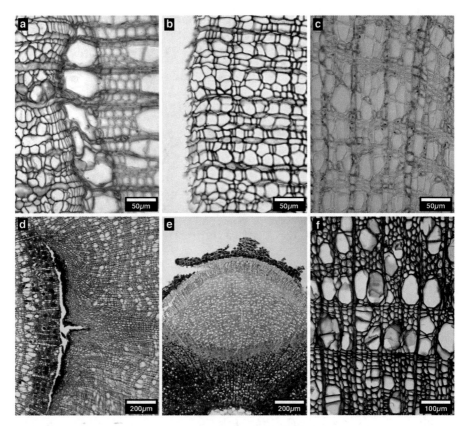

Figure 7. a: Radial growth of *Salix arctica* at the beginning of the growing season, a row of ear-lywood vessels is already lignified but not fully re-enforced by surrounding mechanical tissue. – b: Radial growth of *Rhododendron lapponicum* in the middle of the growing season, three rows of vessels are formed but only the first is lignified, tangential flat latewood cells are missing. – c: Living axial parenchyma and ray cells with nuclei in an 80-year-old part of a stem of *Empetrum nigrum*, stained with Picric-anilinblue. – d: Overgrown wound in a stem of *Arctostaphylos alpina*, living cells inside the wound form a chemical protection zone (phenols) against microbial attack, living cells at the lateral end of the wound form callus cells and overgrew the wound within six years. – e: A barrier zone divides the living and dead zone of a stem of *Cassiope tetragona*, parenchyma cells of the xylem and the phloem outside the barrier zone (light red) represent the living part of the stem, many cells inside the barrier zones are filled with phenols protecting the tissue against decay. – f: Tyloses in the heartwood of *Salix arctica*, interrupting water flow within the xylem. – a–b and d–f stained with safranin-astrablue.

Buizer B, Weijers S, van Bodegom PM, Alsos IG, Eidesen PB, van Breda J, de Korte M, van Rijckevorsel J & Rozema J. 2012. Range shifts and global warming: ecological responses of *Empetrum nigrum* L. to experimental warming at its northern (high Arctic) and southern (Atlantic) geographical range margin. Environm.l Res. Lett. 7: 1–9.

Büntgen U & Schweingruber FH. 2010. Environmental change without climate change? New Phytol. 188: 646–651.

de Witte LC & Stöcklin J. 2010. Longevity of clonal plants: why it matters and how to measure it. Ann. Bot. 106: 859–870.

Elmendorf SC, Henry GHR, Hollister RD, Bjork RG, Boulanger-Lapointe N, Cooper EJ, Cornelissen JHC, Day TA, Dorrepaal E, Elumeeva TG, Gill M, Gould WA, Harte J, Hik DS, Hofgaard A, Johnson DR, Johnstone JF, Jonsdottir IS, Jorgenson JC, Klanderud K, Klein JA, Koh S, Kudo G, Lara M, E. Levesque E, B. Magnusson B, JL. May JL, JA. Mercado-Diaz JA, A. Michelsen A, Molau U, Myers-Smith IH, Oberbauer SF, Onipchenko VG, Rixen C, Schmidt MN, Shaver GR, Spasojevic MJ, Þórhallsdóttir OE, Tolvanen A, Troxler T, Tweedie CE, Villareal S, Wahren C-H, Walker X, Webber PJ, Welker JM & Wipf S. 2012. Plot-scale evidence of tundra vegetation change and links to recent summer warming. Nature Climate Change 2: 453–457.

Gärtner H & Schweingruber FH. 2013. Microscopic preparation techniques for plant stem analysis. Verlag Dr. Kessel, Remagen-Oberwinter. 78 pp.

Good RCR. 1927. The genus *Empetrum* L. J. Linn. Soc. Bot. 47: 489–523.

Greguss P. 1945. Xylotomische Bestimmung der heute lebenden Gymnospermen. Akademiai Kiado, Budapest.

Hallinger M, Manthey M & Wilmking M. 2010. Establishing a missing link: warm summers and winter snow cover promote shrub expansion into alpine tundra in Scandinavia. New Phytol. 186: 890–899.

Hallinger M & Wilmking M. 2011. No change without a cause – why climate change remains the most plausible reason for shrub growth dynamics in Scandinavia. New Phytol. 189: 902–908.

Hultén E. 1968. Flora of Alaska and neighboring territories. Manual of the vascular plants. Stanford University Press, Stanford, California.

Kanngiesser F. 1914. Über Lebensdauer von Zwergsträuchern aus hohen Höhen des Himalaya. Vierteljahrs. Naturf. Ges. Zürich 58: 198–202.

Kaufman DS, Schneider DP, McKay NP, Ammann CM, Bradley RS, Briffa K, Miller GH, Otto-Bliesner BL, Overpeck JT, Vinther BM & Arctic 2k Project Members. 2009. Recent warming reverses long-term Arctic cooling. Science 325: 1236–1239.

Kihlmann A. 1890. Pflanzenbiologische Studien aus Russisch-Lappland. Acta Soc. Fauna Flora Fennoscandia 6: 118.

Kraus G. 1873. Ueber Alter und Wachstumsverhältnisse ostgrönländischer Holzgewächse. Bot. Zeit. 33: 515–518.

Lawrimore JH, Menne MJ, Gleason BE, Williams CN, Wuertz DB, Vose RS & Rennie J. 2011. An overview of the Global Historical Climatology Network monthly mean temperature data set, version 3. J. Geophys. Res. 116 D19 121. DOI:10.1029/2011JD016187.

Macias-Fauria M, Forbes BC, Zetterberg P & Kumpula T. 2012. Eurasian Arctic greening reveals teleconnections and the potential for structurally novel ecosystems. Nature Climate Change 2: 613–618.

McGuire AD, Hayes DJ, Kicklighter DW, Manizza M, Zhuang Q, Chen M, Follows MJ, Gurney KR, McClelland JW, Mellilo JM, Peterson BJ & Prinn RG. 2010. An analysis of the carbon balance of the Arctic Basin from 1997 to 2006. Tellus 62B: 455–474.

Miller HJ. 1975. Anatomical characteristics of some woody plants of the Angmagssalik district of southeast Greenland. Medd. Grønland 198 (6): 1–3. Also in: Meded. Bot. Mus. Utrecht, No. 422.

Molisch H. 1938. The longevity of plants. Science, Lancaster.

Myers-Smith IH, Forbes BC, Wilmking M, Hallinger M, Lantz T, Blok D, Tape KD, Macias-Fauria M, Sass-Klaassen U, Lévesque E, Boudreau S, Ropars P, Hermanutz L, Trant AJ, Collier LS, Weijers S, Rozema J, Rayback SA, Schmidt NM, Schaepman-Strub G, Wipf S, Rixen C, Ménard CB, Venn S, Goetz S, Andreu-Hayles L, Elmendorf S, Ravolainen V, Welker J, Grogan P, Epstein HE & Hik DS. 2011. Shrub expansion in tundra ecosystems: dynamics, impacts and research priorities. Environm. Res. Lett. 6: 045509.

Parsons AN, Welker JM, Wookey PA, Press MC, Callaghan TV & Lee JA. 1994. Growth responses of four sub-Arctic dwarf shrubs to simulated environmental change. J. Ecol. 82: 307–318.

Pauli H, Gottfried M, Dullinger S, Abdaladze O, Akhalkatsi M, Alonso JSB, Coldea G, Dick J, Erschbamer B, Fernández-Calzado R, Ghosn D, Holten JI, Kanka R, Kazakis G, Kollár J, Larsson P, Moiseev P, Moiseev D, Molau I, Molero-Mesa J, Nagy L, Pelino G, Puscas M, Rossi G, Stanisci A, Syverhuset AO, Theurillat JP, Tomaselli M, Unterluggauer P, Villar L, Vittoz P & Grabherr G. 2012. Recent plant diversity changes on Europe's mountain summits. Science 336: 353–355.

Pearson RG. Phillips SJ, Loranty MM, Beck PSA, Damoulas T, Knight SJ & Goetz† SJ. 2013. Shifts in Arctic vegetation and associated feedbacks under climate change. Nature Climate Change. doi: 10.1038/NCLIMATE1858.

Purvis M, Collier D & Walls D. 1964. Laboratory techniques in botany. London, Butterworths.

Rosenthal M. 1904. Über die Ausbildung der Jahrringe an der Grenze des Baumwachstums in den Alpen. Inaug. Diss., Berlin.

Schmidt NM, Baittinger C & Forchhammer MC. 2006. Reconstructing century-long regimes using estimates of high Arctic *Salix arctica* radial growth. Arctic, Antarctic, and Alpine Research 38: 257–262.

Schweingruber FH. 1995. Anatomie europäischer Hölzer. Verlag Paul Haupt, Bern.

Schweingruber FH, Börner A & Schulze ED. 2008. Atlas of woody plants: evolution, structure, and environmental modifications. Springer Verlag, Berlin.

Schweingruber FH. & Poschlod P. 2005. Growth rings in herbs and shrubs: life span, age determination and stem anatomy. Forest Snow and Landscape Research 79: 195–415.

Serreze MC & Francis JA. 2006. The Arctic on the fast track of change. Weather 61: 65–69.

Sturm M, Racine C & Tape K. 2001. Climate change: Increasing shrub abundance in the Arctic. Nature 411: 546–547.

Verbyla D. 2008. The greening and browning of Alaska based on 1982–2003 satellite data. Global Ecology and Biogeography 17: 547–555.

Weijers S, Alsos IG, Eidesen PB, Broekman R, Loonen MJJE & Rozema J. 2012. No divergence in *Cassiope tetragona*: persistence of growth response along a latitudinal temperature gradient and under multi-year experimental warming. Ann. Bot. 110: 653–665.

Wilmking M, Hallinger M, Van Bogaert R, Kyncl T, Babst F, Hahne W, Juday GP, De Luis M, Novak K & Voellm C. 2012. Continuously missing outer rings in woody plants at their distributional margins. Dendrochronologia 30: 213–222.

Accepted: 6 September 2013

IAWA Journal 34 (4), 2013: 498–509

BRILL

DOES GROWTH RHYTHM OF A WIDESPREAD SPECIES CHANGE IN DISTINCT GROWTH SITES?

Monique S. Costa[1],*, Thaís J. de Vasconcellos[1], Claudia F. Barros[2] and Cátia H. Callado[1]

[1]Laboratório de Anatomia Vegetal, Departamento de Biologia Vegetal, Instituto de Biologia Roberto Alcantara Gomes, Universidade do Estado do Rio de Janeiro, Rua São Francisco Xavier 524, Maracanã, 20550-900, Rio de Janeiro, RJ, Brazil
[2]Instituto de Pesquisas Jardim Botânico do Rio de Janeiro, Rua Pacheco Leão 915, Jardim Botânico, 22460-030, Rio de Janeiro, RJ, Brazil
*Corresponding author; e-mail: nique_bio@yahoo.com.br

ABSTRACT

This study is aimed to understand responses in growth rhythm to different climatic conditions of the widespread deciduous species *Cedrela odorata*. Our own research was conducted in Nova Iguaçu. Rio de Janeiro State, Brazil) and compared with literature data from Aripuanã. Mato Grosso State, Brazil), Manaus. Amazonas State, Brazil) and Barinas. Barinas State, Venezuela). Growth periodicity was evaluated through leaf phenological behavior and radial growth. In Nova Iguaçu, leaf phenology was monitored monthly and radial growth was evaluated by cambial histological analysis of samples collected in wet and dry seasons. In the other sites, the authors evaluated the growth rhythm by dendrometer bands. Growth always occurs in the wet season, even when there is no water deficit during the dry season. Thus, the species is considered conservative concerning the maintenance of growth seasonality. Nevertheless, *C. odorata* was able to change its growth period, following local seasonality of its different growth sites. Therefore we suggest caution when performing climate analysis from a chronology using trees that grow in different periods of the year.

Keywords: *Cedrela odorata*, cambial activity, radial growth, phenology, dendrometer bands.

INTRODUCTION

To understand the colonization and survival strategies of tree species in their natural environment, it is important to study their growth dynamics (Callado 2010). This is necessary for the establishment of effective forest management strategies, to recover degraded areas, or for rational and sustainable exploitation (*e.g.*, Worbes 1999, 2002; Rozendaal & Zuidema 2011).

The diversity of growth rhythm responses in tropical species growing in different habitats is still poorly known. Studies indicate that some species retain cambial rhythm from their phytogeographic region of origin when growing under distinct climatic conditions (Fahn 1995; Callado *et al.* 2001, 2013), while other species show local seasonality adaptation (Marcati *et al.* 2006, 2008). These results can directly influence

© International Association of Wood Anatomists, 2013
Published by Koninklijke Brill NV, Leiden

DOI 10.1163/22941932-00000040

dendroclimatological investigations based on trees growing in different sites. Thus, the study of widespread species allows the understanding of how radial growth can vary in their different natural habitats.

Cedrela odorata L. (Meliaceae) (Spanish cedar or Cedro) is naturally distributed from the Mexican Pacific coast and the Mexican Atlantic coast, throughout the Caribbean islands, Yucatan and lowland Central and South America to northern Argentina (Cavers *et al*. 2003). It is recommended and used for the reforestation of degraded areas and produces highly valuable timber (Smith Jr 1965; Cintrón 1990; Lorenzi 1998; Backes & Irgang 2004). Growth periodicity of the species has been studied in different areas (Worbes 1995, 1999; Tomazello-Filho *et al*. 2000; Dünisch & Morais 2002; Dünisch *et al*. 2002, 2003; Brienen & Zuidema 2005, 2006; Brienen *et al*. 2010). In this study we combine our observations with those from other authors who described *Cedrela odorata* growth periodicity in order to investigate its growth rhythm under different climatic conditions.

MATERIAL AND METHODS

Study sites and climate

Our studies were conducted in the Atlantic Rain Forest in the Reserva Biológica do Tinguá, Nova Iguaçu, southeastern Brazil. We evaluated growth periodicity through cambial analysis and leaf phenological behavior. Our data were compared with literature results from *Cedrela odorata* trees (Table 1; Fig. 1) growing in mid-western Brazil Amazon Forest in Aripuanã, Mato Grosso (59° 26' W 10° 9' S) (Dünisch *et al*. 2003); northern Brazil Amazon Forest in Manaus, Amazonia (59° 52' W 3° 8' S) (Dünisch & Morais 2002; Dünisch *et al*. 2002) and south-western Venezuela tropical semi-deciduous forest in Barinas (70° 60' W 7° 30' N) (Worbes 1995, 1999). In these sites, growth periodicity was evaluated by dendrometer bands. In Manaus, growth rhythm was also assessed by using leaf phenological behavior.

In Nova Iguaçu (43° 22' W 22° 34' S) rainfall and temperature data from the study period were obtained from

Figure 1. Study sites where the growth rhythm of *Cedrela odorata* was evaluated.

Table 1. Characteristics of the sites where *Cedrela odorata* growth rhythm was studied. The last column shows the relation between local seasonality and growth periodicity.

Black circles = leaf flushing, white circles = leaf abscission. Largest circles represent the greatest phenophases intensity. Solid lines = radial growth period. Dotted lines = earliest beginning or latest end of the growth period, observed in some trees. In the graphic: dashed line = photoperiod; filled area water excess, corresponding to the wet season; hatched area water deficit, corresponding to the dry season.

(continued on next page →)

SOMAR Meteorologia Ltda and photoperiod data were obtained from Observatório Nacional (Anuário do Observatório Nacional 2004, 2005, 2006, 2008, 2009, 2010). For the comparison of growth seasonality between all the study sites (Nova Iguaçu, Barinas, Aripuanã and Manaus), the results from each site were related to rainfall and temperature data obtained from DIVA-GIS software (Hijmans *et al*. 2005). Water balance diagrams (Walter *et al*. 1975) were constructed for each study site, to show water deficit periods (Table 1). For Nova Iguaçu, water balance diagrams were also built for the study period (Fig. 2).

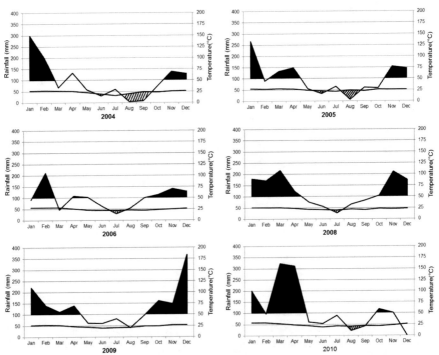

Figure 2. Water balance diagrams of Nova Iguaçu during the study period. Filled area = water excess; hatched area = water deficit.

(Table 1 continued)

Locality	Mean annual rainfall (min–max)	Study period	Growth seasonality and water balance diagrams
			●● ● ○ ○ ○ ....
Nova Iguaçu	1446 mm (25–253 mm)	Cambial analysis: Jul 2004 – May 2005 Mar 2008 – May 2010 Phenological investigation: Jul 2004 – Jun 2006 May 2009 – Jun 2010	
			....
Barinas (10–293 mm) Worbes (1995, 1999)	1741 mm	Dendrometer bands: Apr 1978 – May 1982	
			....
Aripuanã Dünisch *et al.* (2003)	1937 mm (8–321 mm)	Dendrometer bands: Oct 1998 – Nov 2001	
			● ● ○ ○ ○ ○ ....
Manaus Dünisch *et al.* (2002) Dünisch & Morais (2002)	2148 mm (59–291 mm)	Dendrometer bands: Jan 1995 – Dec 1998 Phenological investigation: Dec 1997 – Feb 1999	

Trees studied

In our experiment, ten straight-boled adult trees of *C. odorata* were selected. Trees occupied the emergent layer of this forest and were 35.2 (± 3.6) m high and 69.6 (± 6.7) cm in diameter at breast height.

Phenological observations

The vegetative phenology was monitored monthly from July 2004 to June 2006 and from May 2009 to June 2010, through direct observation of the canopy with binoculars. The vegetative phenophases monitored were: leaf flushing, leaf abscission and the presence of young, adult or senescent leaves. The intensity of each phenological event was scored as indices 0 to 4, corresponding to 1–25%, 26–50%, 51–75%, or 76–100% of leaves present, respectively (Fournier 1974). Fournier's percent index of intensity was calculated by summing 0 to 4 indices and dividing it by the maximum possible number. number of trees multiplied by 4). The obtained value is then multiplied by 100 to become a percent value (Fournier 1974). The synchrony of the vegetative phenophases was calculated according to the index proposed by Augspurger (1983). Based on this index, phenological events were classified according to the level of synchrony. absence of synchrony: <20% of the trees in a given phenophase; low synchrony: 20–60% of the trees in the phenophase; high synchrony: >60% of the trees in the phenophase (Bencke & Morellato 2002).

Cambial activity

For cambial analysis, stem samples were obtained with a saw, chisel and hammer, at a distance of about 1.30 m above the ground. These samples were obtained in July and October of 2004; February and May of 2005; August of 2008; March, May, August and November of 2009; February and May of 2010, representing the wet and dry seasons of the year (Table 2).

Table 2. Sampling periods and characteristics of the cambial zone of *Cedrela odorata* from Atlantic Rain Forest.

Sampling/ Study year	Dry season (Winter)		Wet season (Spring)		Wet season (Summer)		Dry season (Autumn)	
2004–2005	Jul (n = 4)	D	Oct (n = 4)	A	Feb (n = 4)	A	May (n = 4)	D
2008–2009	Aug (n = 4)	D	—		Mar (n = 9)	A	May (n = 9)	D
2009–2010	Aug (n = 9)	D	Nov (n = 9)	A	Feb (n = 9)	A	May (n = 9)	D

n = number of sampled trees; D = dormant; A = active.

In order to optimize infiltration processes, samples were reduced to 1-cm^3 specimens, which were fixed in CRAF III. Sass 1958) and dehydrated in ethanol. Two specimens of each tree and for each sampling, were randomly chosen to be embedded in resin glycol methacrylate (Feder & O'Brien 1968) and sectioned with a rotary microtome at an average thickness of 5 μm, in transverse and radial-longitudinal sections. Histological sections were stained with toluidine blue O (Johansen 1940) or with the fluorochromes auramine O and aniline blue (Barros & Miguens 1998; Ruzin 1999). The number of cell

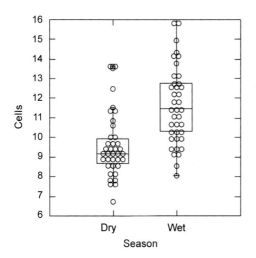

Figure 3. Box-plots showing distributions of the number of cell layers in the cambial zone of *Cedrela odorata* in wet and dry seasons.

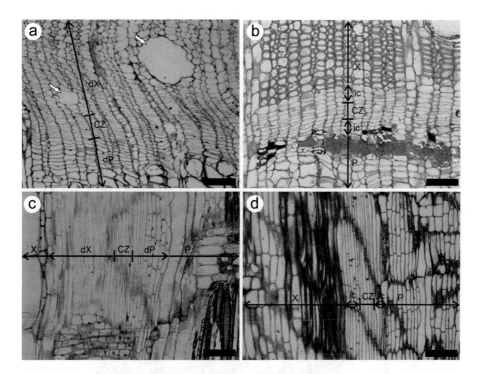

Figure 4. Cambial zone of *Cedrela odorata* in the different growing stages. – a: Cross section of active period (November 2004). The white arrows indicate the differentiation of a vessel element in the secondary xylem adjacent to the cambial zone. – b: Cross section of dormancy period (May 2010). – c: Radial-longitudinal section of active period (November 2009). – d: Radial-longitudinal section of dormancy period (May 2010). – X = secondary xylem; dX = differentiating secondary xylem; CZ = cambial zone; dP = differentiating secondary phloem; P = secondary phloem; ic = immature cells. — Scale bar = 50 μm.

layers in the cambial zone was counted using light microscopy. Statistical comparisons among the number of cell layers in the cambial zone between wet and dry periods were made with a t-test (Zar 1999).

RESULTS

Growth periodicity in the Atlantic Rain Forest

Cedrela odorata showed annual decid-uous behavior, with highly synchronous leaf abscission (synchrony index of 0.87) beginning in April / May and persistent until July (Table 1). Leaf flushing occurred between July and September, but mainly in August (Table 1), and was also classified as highly synchronous (synchrony index of 0.84). From October on, the trees had adult leaves, which became senescent in March /April.

Radial growth seasonality was evident from histological analysis of the cambial zone and the adjacent vascular tissues. The number of cell layers in the cambial zone differed significantly between wet and dry seasons (t = -5.343; p < 0,001– see Figure 3). In the dry season, the cambial zone showed dormancy traits, with radially expanded cells showing thick walls. Close to the cambial zone, fully differentiated secondary xylem with lignified cell walls and secondary phloem with callose deposition in sieve tube elements were observed (Fig. 4b, d and 5). The last differentiated xylem cells row were thick-walled fibers, corresponding to the formation of latewood (Fig. 6b, d). Thus, the parenchyma bands in the limit of growth rings were classified as initial, since they do not differentiate before cambial dormancy.

In the wet season, the cambial zone was active (Table 1 & 2), with narrow cells and thin walls, indicating intensive cell division (Fig. 4a, c). We observed a great increase in the number of cell layers in the cambial zone in this period. Furthermore, the secondary xylem presented many cell layers going through elongation and differentiation

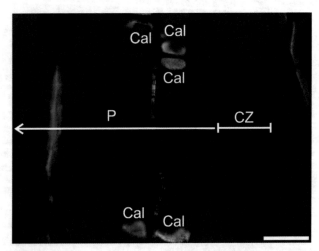

Figure 5. Radial-longitudinal section of secondary phloem of *Cedrela odorata* during the dormancy period (May 2010). Note callose deposition in sieve tube elements adjacent to the cambial zone, evidenced by aniline blue fluorescence. – Cal = sieve tube elements with callose; P = secondary phloem; CZ = cambial zone. — Scale bar = 50 μm.

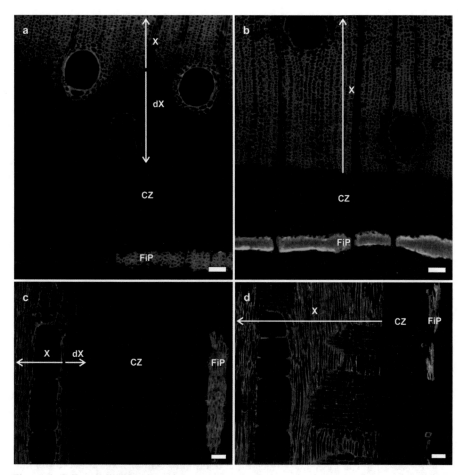

Figure 6. Cell wall lignification of secondary xylem adjacent to the cambial zone of *Cedrela odorata*, evidenced by Auramine O fluorescence. – a: Cross section of active period (February 2010). – b: Cross section of dormancy period (May 2010). – c: Radial-longitudinal section of active period (February 2010). – d: Radial-longitudinal section of dormancy period (May 2010). – FiP = secondary phloem fibers; CZ = cambial zone; dX = differentiating secondary xylem; X = secondary xylem. — Scale bar = 100 μm.

phases (Fig. 4a, c). The different stages of xylem cell wall lignification could be better visualized by the application of the fluorochrome auramine O, which showed a gradient of lignification toward the cambial zone (Fig. 6a, c). During the active phase, we did not find callose deposition in sieve tube elements close to the cambial zone.

Cambial dormancy was observed from May to August. In October, the cambial zone was very active, with many cell layers (13.50 ± 1.31 – Table 2), and a great number of differentiating xylem and phloem cells. Thus, we inferred that cambial activity begins in September and goes on until March/April. This growth pattern was observed even when there was no water deficit in the dry season, as observed in 2009 (Fig. 2).

Both cambial dormancy and leaf abscission began in the end of the wet season, coinciding to the period when the days become shorter than the nights and rainfall starts to decrease (Table 1). This dormant stage remained during the dry season, but leaf flushing began before the rainfall increased, coinciding with the lengthening of the photoperiod.

Growth seasonality in different study sites

The comparison of the growth seasonality of *Cedrela odorata* in the Atlantic Rain Forest, Amazon-Aripuanã, Amazon-Manaus and the Tropical semi-deciduous forest showed that its growth always occurs in the wet season, even if the rainy periods of each region occur in different months (Table 1). In Amazon-Manaus, growth seasonality did not follow climatic means obtained from DIVA-GIS software (Table 1); however, both cambial growth and the presence of leaves coincided with the wet season (January to July) reported by Dünisch *et al.* (2002) and Dünisch and Morais (2002) for their study period.

DISCUSSION

Despite the differences between growth period among the different sites and the different annual conditions, there is a clear seasonality of *Cedrela odorata* growth rhythm, determined by leaf abscission and cambial dormancy in the drier period of the year, even when there is no water deficit in the dry period, like in 2009 in Nova Iguaçu. Furthermore, the correlation of tree growth in Manaus-Amazonia with rainfall seasonality of the actual study period (Dünisch *et al.* 2002; Dünisch & Morais 2002) – and not with DIVA-GIS historical means – suggests that the species is able to respond to annual changes, since the driest season reported for the study years was remarkably different from the historical means. Therefore, the reduction of rainfall seems to be the factor which leads to growth seasonality of this species.

In the Atlantic Rain Forest, besides the reduction of rainfall, we observed that the first changes in leaf phenology and cambial dormancy coincide also with a reduction of photoperiod. This agrees with Borchert and Rivera (2001) who reported that photoperiod variation signals the onset of wet or dry seasons.

The interrelation between signaling of the driest period, subsequent leaf abscission and cambial dormancy becomes clear when we compare our results with other studies (*e.g.*, Callado *et al.* 2001; Rivera *et al.* 2002). In both sites where *C. odorata* leaf phenology was assessed, Atlantic Rain Forest and Manaus-Amazonia, cambial dormancy was observed soon after leaf abscission, during the driest season. Leaf loss resulting from seasonal reduction of rainfall and the consequent termination of photoassimilate production leads to cambial dormancy, since investing in radial growth is not a priority of the plant when its metabolism is reduced (Savidge 2000).

The association between leaf behavior and responses in formation and physiology of conducting tissues are also observed at other scales. According to the literature, differentiation of latewood fibers is associated with the production of gibberellins by mature leaves (*e.g.* Aloni 2007), and this is consistent with the formation of latewood fibers at the end of the growth period observed in *C. odorata*. Also, dormancy callose observed during deciduous period agrees with the report of Aloni *et al.* (1990), who described that immediately after leaf abscission the effect of cytokinin would predominate over auxin, leading to heavy callose production.

We point out that in Amazon-Aripuanã, Amazon-Manaus and the Tropical semideciduous forest the growth period continued until the first month of the driest season. In the Atlantic Rain Forest, we observed that the growth period stops one or two months before the dry season.

The maintenance of the deciduous habit and cambial dormancy of *C. odorata* even in a site where water is not a limiting factor, like Amazon-Manaus, indicates that the species retains a typical behavior of a seasonal dry environment. Muellner *et al.* (2010) stated that the deciduous habit of *Cedrela* and other morphological adaptations, like shoot apices protected by a cluster of bud scales and capsular fruits with dry, winged and wind-dispersed seeds of extended viability, point to an evolutionary history in dry habitats. This suggests a conservative nature of *C. odorata* concerning the maintenance of growth seasonality. On the other hand, the species showed the ability to change its growth period, following local seasonality of its different growth sites, which differs from the findings of Callado *et al.* (2001) studying the periodicity of *Symphonia globulifera* growth rings in the Atlantic Rain Forest. The authors reported that latewood formation did not coincide with phenophases or with local climatic variations, but seemed to repeat the rhythm observed for this species in the Amazon floodplains, corresponding to the aquatic phase of this ecosystem.

The fact that *C. odorata* is a widespread species that follows local climate and has annual growth rings (Worbes 1995, 1999; Botosso *et al.* 2000; Tomazello-Filho *et al.* 2000; Dünisch *et al.* 2002, 2003; Brienen & Zuidema 2005, 2006) could recommend it for the construction of a tree ring chronology for Latin America. However, its growth period occurs in different periods of the year, depending on rainfall seasonality in each site. We therefore suggest caution against climate analysis through a chronology with trees growing in different periods of the year, since the species growth period is used to test climatic correlations. The next step should be to test a chronology among the sites where radial growth occurs in the same period of the year. Finally, this relation between growth and local seasonality should also be investigated for other widespread species when building chronologies with trees from different sites.

ACKNOWLEDGEMENTS

We thank the Universidade do Estado do Rio de Janeiro. UERJ), Fundação Carlos Chagas Filho de Amparo à Pesquisa do Rio de Janeiro. FAPERJ), Conselho Nacional de Desenvolvimento Científico e Tecnológico. CNPq) and Coordenação de Aperfeiçoamento de Pessoal de Nível Superior. CAPES) for fellowships, funding and research grants; Walter da Silva for his invaluable support in all field works; João Ferreira Alves Junior and Eduardo Gama Mendes de Moraes for their help in data achievement at the beginning of this study; Maxmira de Souza Arêdes for support in samples processing and the journal editor and reviewers for their valuable comments and suggestions.This paper was derived from the master dissertation of the first author.

REFERENCES

Aloni R. 2007. Phytohormonal mechanisms that control wood quality formation in young and mature trees. In: Entwistle K, Harris P & Walker J. (eds.), The compromised wood workshop: 1–22. The Wood Technology Research Centre, University of Canterbury, Christchurch, New Zealand.

Aloni R, Baum SF & Peterson CA. 1990. The role of cytokinin in sieve tube regeneration and callose production in wounded coleus internodes. Plant Physiol. 93: 982–989.

Anuário do Observatório Nacional. 2004–2006, 2008–2010. Continuação de Efemérides Astronômicas do Observatório Nacional, vol. 120–122, 124–126. Observatório Nacional, Rio de Janeiro.

Augspurger CK. 1983. Phenology, flowering synchrony and fruitset of six neotropical shrubs. Biotropica 15: 257–267.

Backes P & Irgang B. 2004. Mata Atlântica: as árvores e a paisagem. Paisagem do Sul, Porto Alegre. 393 pp.

Barros CF & Miguens FC. 1998. Ultrastructure of the epidermal cells of *Beilschmiedia rigida* (Mez) Kosterm. (Lauraceae). Acta Microscopica 6: 451–461.

Bencke CSC & Morellato LPC. 2002. Comparação de dois métodos de avaliação da fenologia de plantas, sua interpretação e representação. Rev. Bras. Bot. 25: 269–275.

Borchert R & Rivera G. 2001. Photoperiodic control of seasonal development and dormancy in tropical stem succulent trees. Tree Physiol. 21: 213–221.

Botosso PC, Vetter RE & Tomazello-Filho M. 2000. Periodicidade e taxa de crescimento de árvores de cedro (Cedrela odorata L., Meliaceae), jacareúba (Calophyllum angulare A.C., Clusiaceae) e muirapiranga (Eperua bijuga Mart. ex Benth., Leg. Caesalpinioideae). In: Roig FA (ed.), Dendrocronologia em America Latina: 357–379. EDIUNC, Mendoza.

Brienen RJW & Zuidema PA. 2005. Relating tree growth to rainfall in Bolivian rain forests: a test for six species using tree ring analysis. Oecologia 146: 1–12.

Brienen RJW & Zuidema PA. 2006. Lifetime growth patterns and ages of Bolivian rain forest trees obtained by tree ring analysis. J. Ecol. 94: 481–493.

Brienen RJW, Zuidema PA & Martínez-Ramos M. 2010. Attaining the canopy in dry and moist tropical forests: strong differences in tree growth trajectories reflect variation in growing conditions. Oecologia 163: 485–496.

Callado CH. 2010. Os anéis de crescimento no estudo da dinâmica populacional na Floresta Atlântica. In: Absy ML, Matos FDA & Amaral IL (eds.), Diversidade Vegetal Brasileira: conhecimento, conservação e uso: 227–231. Sociedade Botânica do Brasil, Manaus.

Callado CH, Roig FA, Tomazello-Filho M & Barros CF. 2013. Cambial growth periodicity studies of South American woody species – A review. IAWA J. 34: 213–230.

Callado CH, Silva Neto SJ, Scarano FR & Costa CG. 2001. Periodicity of growth rings in some flood-prone trees of the Atlantic Rain Forest in Rio de Janeiro, Brazil. Trees 15: 492–497.

Cavers S, Navarro C & Lowe AJ. 2003. Chloroplast DNA phylogeography reveals colonisation history of a Neotropical tree, *Cedrela odorata* L., in Mesoamerica. Molec. Ecol. 12: 1451–1460.

Cintron BB. 1990. *Cedrela odorata* L. Cedro Hembra, Spanish Cedar, Meliaceae. Mahogany family: 250–257. In: Silvics of North America. Hardwoods. v.2. Washington, DC, Agric. Handbook, USDA.

Dünisch O, Bauch J & Gasparotto L. 2002. Formation of increment zones and intra-annual growth dynamics in the xylem of *Swietenia macrophylla*, *Carapa guianensis*, and *Cedrela odorata* (Meliaceae). IAWA J. 23: 101–119.

Dünisch O, Montóia VR & Bauch J. 2003. Dendroecological investigations on *Swietenia macrophylla* King and *Cedrela odorata* L. (Meliaceae) in the central Amazon. Trees 17: 244–250.

Dünisch O & Morais RR. 2002. Regulation of xylem sap flow in an evergreen, a semi-deciduous, and a deciduous Meliaceae species from the Amazon. Trees 16: 404–416.

Fahn A. 1995. Seasonal cambial activity and phytogeographic origin of woody plants: a hypothesis. Isr. J. Plant Sci. 43: 69–75.

Feder N & O'Brien TP. 1968. Plant microtechnique: some principles and new methods. Amer. J. Bot. 55: 123–142 .

Fournier LA. 1974. Un método cuantitativo para la medición de características fenológicas en árboles. Turrialba 24: 422–423.

Hijmans RJ, Cameron SE, Parra JL, Jones PG & Jarvis A. 2005. Very high resolution interpolated climate surfaces for global land áreas. Int. J. Climatol. 25: 1965–1978.

Johansen DA. 1940. Plant microtechnique. McGraw-Hill, New York.

Lorenzi H. 1998. Árvores Brasileiras. Manual de Identificação e Cultivo de Plantas Arbóreas do Brasil, vol. 2. Plantarum, Nova Odessa. 231 pp.

Marcati CR, Angyalossy V & Evert RF. 2006. Seasonal variation in wood formation of *Cedrela fissilis* (Meliaceae). IAWA J. 27: 199–211.

Marcati CR, Milanez CRD & Machado SR. 2008. Seasonal development of secondary xylem and phloem in *Schizolobium parahyba* (Vell.) Blake (Leguminosae: Caesalpinioideae). Trees 22: 3–12.

Muellner AN, Pennington TD, Koecke AV & Renner SS. 2010. Biogeography of *Cedrela* (Meliaceae, Sapindales) in Central and South America. Amer. J. Bot. 97: 511–518.

Rivera G, Elliott S, Caldas LS, Nicolossi G, Coradin VT & Borchert R. 2002. Increasing daylength induces spring flushing of tropical dry forest trees in the absence of rain. Trees 16: 445–456.

Rozendaal DMA & Zuidemal PA. 2011. Dendroecology in the tropics: a review. Trees 25: 3–16.

Ruzin SE. 1999. Plant microtechnique and microscopy. Oxford University Press, Cambridge.

Sass JE. 1958. Elements of botanical microtechnique. Vol. 2. McGraw-Hill Book Company, New York.

Savidge RA. 2000. Intrinsic regulation of cambial growth. J. Plant Growth Regul. 20: 52–77.

Smith Jr CE. 1965. Flora of Panama. Part VI, Family 92, Meliaceae. Ann. Missouri Bot. Gard. 52: 55–79.

Tomazello-Filho M, Botosso PC & Lisi CS. 2000. Potencialidade da família Meliaceae para dendrocronologia em regiões tropicais e subtropicais. In: Roig FA (org.), Dendrocronología en América Latina: 381–434. EDIUNC, Mendoza.

Walter H, Harnickell E & Mueller-Dombois D. 1975. Climate-diagram maps. Springer Verlag, Berlin.

Worbes M. 1995. How to measure growth dynamics in tropical trees – A review. IAWA J. 16: 337–351.

Worbes M. 1999. Annual growth rings, rainfall-dependent growth and long-term growth patterns of tropical trees from the Caparo Forest Reserve in Venezuela. J. Ecol. 87: 391–403.

Worbes M. 2002. One hundred years of tree-ring research in the tropics – a brief history and an outlook to future challenges. Dendrochronologia 20: 217–231.

Zar JH. 1999. Biostatistical analysis. Prentice Hall, New Jersey.

Accepted: 2 September 2013